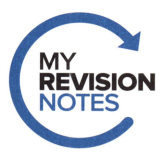

KV-577-433

D

Please return/renew this item by the last date shown. Books may also be renewed by phone or internet.

🖥 www.rbwm.gov.uk/home/leisure-and-culture/libraries

☎ 01628 796969 (library hours)

☎ 0303 123 0035 (24 hours)

Royal Borough
of Windsor &
Maidenhead

www.rbwm.gov.uk

HODDER
EDUCATION
AN HACHETTE UK COMPANY

The Publishers would like to thank the following for permission to reproduce copyright material.

Acknowledgements

AQA material is reproduced by permission of AQA.

Page 32: Image adapted from *Reference Ranges for Spirometry Across All Ages – A New Approach.* Stanojevic, S. et al, Am. J. Respir. Crit. Care Med. 2008; 177(3): 253–260.

Page 92: Definitions available at https://filestore.aqa.org.uk/resources/science/AQA-1775-GOT.PDF

Pages 124–126: reproduced with the permission of World Nuclear Association, https://www.world-nuclear.org/information-library/safety-and-security/safety-of-plants/chernobyl-accident.aspx

Pages 126–127: © 2020 Ripley Entertainment Inc.

Every effort has been made to trace all copyright holders, but if any have been inadvertently overlooked, the Publishers will be pleased to make the necessary arrangements at the first opportunity.

Although every effort has been made to ensure that website addresses are correct at time of going to press, Hodder Education cannot be held responsible for the content of any website mentioned in this book. It is sometimes possible to find a relocated web page by typing in the address of the home page for a website in the URL window of your browser.

Hachette UK's policy is to use papers that are natural, renewable and recyclable products and made from wood grown in well-managed forests and other controlled sources. The logging and manufacturing processes are expected to conform to the environmental regulations of the country of origin.

Orders: please contact Hachette UK Distribution, Hely Hutchinson Centre, Milton Road, Didcot, Oxfordshire, OX11 7HH. Telephone: +44 (0)1235 827827. Email education@hachette.co.uk. Lines are open from 9 a.m. to 5 p.m., Monday to Friday. You can also order through our website: www.hoddereducation.co.uk

ISBN: 978 1 3983 1762 8

© Jeremy Pollard and Adrian Schmit 2021

First published in 2021 by
Hodder Education,
An Hachette UK Company
Carmelite House
50 Victoria Embankment
London EC4Y 0DZ

www.hoddereducation.co.uk

Impression number   10 9 8 7 6 5 4 3 2 1

Year   2025 2024 2023 2022 2021

Cover photo © PhotoEdit – stock.adobe.com

Typeset by Integra Software Services Pvt. Ltd., Pondicherry, India

Printed in Spain

A catalogue record for this title is available from the British Library.

Get the most from this book

Everyone has to develop their own revision strategy, but it is essential to review your work, learn it and test your understanding. These Revision Notes will help you to do that in a planned way, topic by topic. Use this book as the cornerstone of your revision and don't hesitate to write in it — personalise your notes and check your progress by ticking off each section as you revise.

Tick to track your progress

Use the revision planner on pages 4–6 to plan your revision, topic by topic. Tick each box when you have:

+ revised and understood a topic
+ tested yourself
+ practised the exam questions and gone online to check your answers and complete the quick quizzes.

You can also keep track of your revision by ticking off each topic heading in the book. You may find it helpful to add your own notes as you work through each topic.

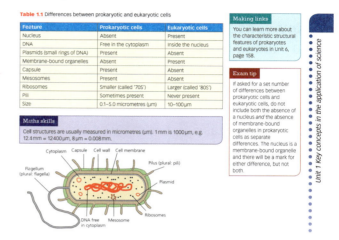

Table 1.1 Differences between prokaryotic and eukaryotic cells

Feature	Prokaryotic cells	Eukaryotic cells
Nucleus	Absent	Present
DNA	Free in the cytoplasm	Inside the nucleus
Plasmids (small rings of DNA)	Present	Absent
Membrane-bound organelles	Absent	Present
Capsule	Present	Absent
Mesosomes	Present	Absent
Ribosomes	Smaller (called '70S')	Larger (called '80S')
Pili	Sometimes present	Never present
Size	0.1–5.0 micrometres (µm)	10–100 µm

Maths skills

Cell structures are usually measured in micrometres (µm). 1 mm is 1000 µm, e.g. 12.4 mm = 12400 µm, 8 µm = 0.008 mm.

Making links

You can learn more about the characteristic structural features of prokaryotes and eukaryotes in Unit 6, page 158.

Exam tip

If asked for a set number of differences between prokaryotic and eukaryotic cells, do not include both the absence of a nucleus *and* the absence of membrane-bound organelles in prokaryotic cells as separate differences. The nucleus is a membrane-bound organelle and there will be a mark for either difference, but not both.

Features to help you succeed

Exam tips

Expert tips are given throughout the book to help you polish your exam technique and maximise your chances in the exam.

Now test yourself

These short, knowledge-based questions provide the first step in testing your learning. Answers are provided at the back of the book.

Definitions and key words

Clear, concise definitions of essential key terms are provided where the terms first appear.

Key words from the specification are highlighted in bold throughout the book.

Making links

This feature identifies specific connections between topics and tells you how revising these will aid your exam answers.

Exam practice

Practice exam questions are provided for each topic. Use them to consolidate your revision and practise your exam skills.

Summaries

The summaries provide a quick-check bullet list for each topic.

Online

Go online to check your answers to the exam questions and try out the extra quick quizzes at **www.hoddereducation.co.uk/myrevisionnotesdownloads**

Worked examples

These take you step-by-step through a problem to help you understand how to answer similar questions in future.

Maths skills

These highlight some of the key mathematical skills you will need to know to do well in your exam. They explain how you should approach a calculation before giving you some practice question so you can practise what you've learnt.

My Revision Notes: AQA Applied Science Suitable for Level 3 and Level 3 Extended Certificates

My Revision Planner

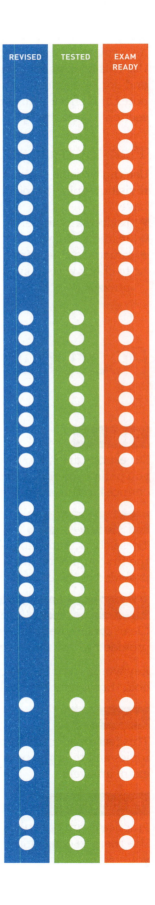

REVISED TESTED EXAM READY

Check your understanding and progress at **www.hoddereducation.co.uk/myrevisionnotes**

REVISED   TESTED   EXAM READY

My Revision Planner

REVISED   TESTED   EXAM READY

Check your understanding and progress at **www.hoddereducation.co.uk/myrevisionnotes**

Countdown to my exams

6–8 weeks to go

✚ Start by looking at the specification — make sure you know exactly what material you need to revise and the style of the examination. Use the revision planner on pages 4–6 to familiarise yourself with the topics.

✚ Organise your notes, making sure you have covered everything on the specification. The revision planner will help you to group your notes into topics.

✚ Work out a realistic revision plan including time for relaxation. Set aside days and times for all the subjects that you need to study, and stick to your timetable.

✚ Set yourself sensible targets. Break down your revision into focused sessions of around 40 minutes, divided by breaks. These Revision Notes organise the basic facts into short, memorable sections to make revising easier.

REVISED ⬤

2–6 weeks to go

✚ Read through the relevant sections of this book and refer to the exam tips, summaries and key terms. Tick off the topics as you feel confident about them. Highlight those topics you find difficult and look at them again in detail.

✚ Test your understanding of each topic by working through the 'Now test yourself' questions in the book. Look up the answers at the back of the book.

✚ Make a note of any problem areas as you revise, and ask your teacher to go over these in class.

✚ Look at past papers. They are one of the best ways to revise and practise your exam skills. Write or prepare planned answers to the exam practice questions provided in this book. Check your answers online and try out the extra quick quizzes at **www.hoddereducation.co.uk/myrevisionnotesdownloads**

✚ Try out different revision methods. For example, you can make notes using mind maps, spider diagrams or flash cards.

✚ Track your progress using the revision planner and give yourself a reward when you have achieved your target.

REVISED ⬤

One week to go

✚ Try to fit in at least one more timed practice of an entire past paper and seek feedback from your teacher, comparing your work closely with the mark scheme.

✚ Check the revision planner to make sure you haven't missed out any topics. Brush up on any areas you find difficult by talking them over with a friend or getting help from your teacher.

✚ Attend any revision classes put on by your teacher. Remember, he or she is an expert at preparing people for examinations.

REVISED ⬤

The day before the examination

✚ Flick through these Revision Notes for useful reminders, for example, examiners' tips, examiners' summaries and key terms.

✚ Check the time and place of your examination.

✚ Make sure you have everything you need — extra pens and pencils, highlighter pen, tissues, a watch, bottled water, sweets.

✚ Allow some time to relax and have an early night to ensure you are fresh and alert for the examinations.

REVISED ⬤

My exams – when and where

....
....
....
....
....
....
....
....
....

My Revision Notes: AQA Applied Science Suitable for Level 3 and Level 3 Extended Certificates

Exam breakdown

These pages outline how AQA Level 3 Applied Science will be examined. There are two qualifications available following this specification – Certificate and Extended Certificate. This book covers both, but you should check which qualification you are taking with your teacher, so that you only learn the material relevant to you.

The qualifications break down as follows.

Level 3 Certificate in Applied Science consists of **three** mandatory units:

Unit/Chapter	Unit title	Assessment type	Unit weighting (%)
1	Key concepts in science	Written examination	33.3
2	Applied experimental techniques	Portfolio	33.3
3	Science in the modern world	Written examination with pre-release material	33.3

Level 3 Extended Certificate in Applied Science consists of **five** mandatory units and **one optional unit** from a choice of three. Again, it is worth checking which optional unit you are studying:

Unit/Chapter	Unit title	Assessment type	Unit weighting (%)
Mandatory			
1	Key concepts in science	Written examination	16.6
2	Applied experimental techniques	Portfolio	16.6
3	Science in the modern world	Written examination with pre-release material	16.6
4	The human body	Written examination	16.6
5	Investigating science	Portfolio	16.6
Optional			
6a	Microbiology	Portfolio	16.6
6b	Medical physics	Portfolio	16.6
6c	Organic chemistry	Portfolio	16.6

Assessment details

REVISED

Each unit may be assessed in January and June of each year – check when you are likely to be assessed so you can ensure you're prepared well in advance. Below is a quick summary, but you can find more information including past papers, mark schemes and Examiner's Reports on the AQA website.

+ **Unit 1 Key concepts in science.**
  This is a written examination that takes place over 1 hour and 30 minutes. It has three sections:
  + A Applications of biology
  + B Applications of chemistry
  + C Applications of physics

  Each section will be allocated 20 marks and will consist of short-answer questions set in applied contexts.

  You will be given a Formulae Sheet containing all the mathematical formulae required for Unit 1 **plus** a copy of the Periodic Table.

- **Unit 2 Applied experimental techniques**.
  This is a portfolio of six experimental reports based on the following experiments:
  + 1(a) Rate of respiration
  + 1(b) Light-dependent reaction in photosynthesis (the Hill reaction)
  + 2(a) Volumetric analysis
  + 2(b) Colorimetric analysis
  + 3(a) Resistivity
  + 3(b) Specific heat capacity

  plus written risk assessments for **one** applied experimental technique from **each** of biology, chemistry and physics (three in total).

- **Unit 3 Science in the modern world**.
  This is a written examination (with pre-released materials) that takes place over 1 hour and 30 minutes and has an allocation of 60 marks. It is in two parts:
  + **Section A** – short and extended answer questions (usually) based on four 'sources' published in the pre-release material, provided to centres approximately 2 months before the examination
  + **Section B** – short and extended answer questions not based on the pre-release material, although they may or may not be based on the same topical scientific issue. Questions will generally include some form of analysis of data presented in the examination paper.

- **Unit 4 The human body**.
  This is a written examination, taking place over 1 hour and 30 minutes. It has an allocation of 60 marks, and comprises short-answer questions.

- **Unit 5 Investigating science**.
  This is a **portfolio** of reports (which can be in a variety of formats) based on **one** practical investigation approved by AQA – either from their pre-approved list or approved through your centre.

- Unit 6 is in three parts:
  + **6a Microbiology** is a **portfolio** of reports (which can be in a variety of formats):
    - identifying the main groups of microorganisms in terms of their structure
    - using aseptic techniques to safely culture microorganisms
    - using practical techniques to investigate factors that affect the growth of microorganisms
    - identifying the use of microorganisms in biotechnological industries.
  + **6b Medical physics** is a **portfolio** of reports (which can be in a variety of formats):
    - understanding imaging methods
    - understanding radiotherapy techniques and the use of radioactive tracers
    - demonstrating the ability to work with radioisotopes in the laboratory
    - understanding the medical uses of optical fibres and lasers.
  + **6c Organic chemistry** is a **portfolio** of reports (which can be in a variety of formats):
    - identifying molecular structure, functional groups and isomerism
    - understanding reactions of functional groups
    - preparing organic compounds.

Unit 1 Key concepts in the application of science

Key concepts in the application of biology

Introduction

In this section you will learn that cells are the basic units of life, and that a knowledge of their structure and function is vital for the understanding of diseases, fertility, growth and development as well as the effect of a variety of environmental chemicals, including drugs.

You will also learn about the heart. Heart and circulatory system disease cause more than a quarter of the deaths in the UK.

Then you will learn that life consists of a series of co-ordinated chemical reactions. For the chemical systems to work efficiently, certain factors must be kept at a constant or near constant level. The maintenance of this steady state is called homeostasis. Factors controlled include: body temperature; the concentration of body fluids; and blood sugar and pH levels.

You will also discover that respiration is the process which provides all cells with the energy they need for living processes. The energy comes from food materials, but its efficient extraction requires oxygen and, in many animals, including humans, oxygen enters the body by the process of breathing.

Finally, you will learn that the world's food supply depends upon the process of photosynthesis, by which plants make food. Animals rely on plants for their food supply, either directly or indirectly.

> **Making links**
>
> You can learn more about the roles and responsibilities of biomedical scientists, pharmacologists, biochemists, environmental scientists, research scientists and sport and exercise scientists in Unit 3. See page 122.

Cell structure

Cell types have differences in ultrastructure

The development of high-magnification electron microscopes has led to discoveries about the ultrastructure of cells.

Cells come in two types:
1 prokaryotic
2 eukaryotic.

Prokaryotic cells are those found in bacteria and blue-green algae, while eukaryotic cells make up animals, plants and fungi. Both types of cell have cytoplasm and a cell membrane. The differences between the two types of cell are shown in Figure 1.1 and Table 1.1.

Eukaryotic Cells which contain membrane-bound organelles and a nucleus.

Prokaryotic Prokaryotic cells are simple cells that do not have a true nucleus or other membrane-bound organelles.

Ultrastructure The fine structure of cells which is only visible with electron microscopes.

Table 1.1 Differences between prokaryotic and eukaryotic cells

Feature	Prokaryotic cells	Eukaryotic cells
Nucleus	Absent	Present
DNA	Free in the cytoplasm	Inside the nucleus
Plasmids (small rings of DNA)	Present	Absent
Membrane-bound organelles	Absent	Present
Capsule	Present	Absent
Mesosomes	Present	Absent
Ribosomes	Smaller (called '70S')	Larger (called '80S')
Pili	Sometimes present	Never present
Size	0.1–5.0 micrometres (µm)	10–100 µm

Making links

You can learn more about the characteristic structural features of prokaryotes and eukaryotes in Unit 6, page 158.

Exam tip

If asked for a set number of differences between prokaryotic cells and eukaryotic cells, do not include both the absence of a nucleus *and* the absence of membrane-bound organelles in prokaryotic cells as separate differences. The nucleus is a membrane-bound organelle and there will be a mark for either difference, but not both.

Maths skills

Cell structures are usually measured in micrometres (µm). 1 mm is 1000 µm, e.g. 12.4 mm = 12 400 µm, 8 µm = 0.008 mm.

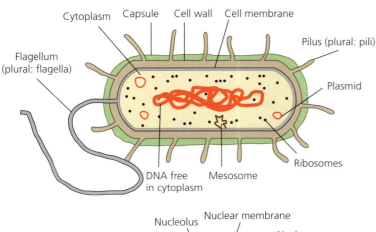

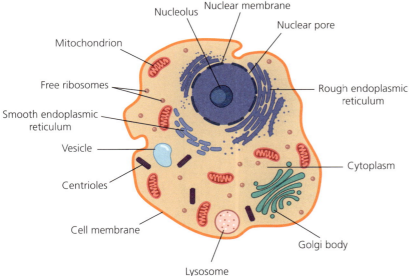

Figure 1.1 Structure of a typical prokaryotic and eukaryotic cell. The eukaryotic cell show here is an animal cell

Note that the eukaryotic cell shown in Figure 1.1 is an animal cell. Plant cells have a cell wall and sometimes chloroplasts, and no centrioles.

Ultrastructure and functions of cell organelles

Both prokaryotic and eukaryotic cells have specialised organelles inside them, each of which carries out a particular function. The functions of each organelle are shown in Table 1.2.

Organelle Structure found inside a cell which has a specific function.

11

Table 1.2 Functions of organelles

Organelle	Function	Notes
Nucleus	Contains the DNA which controls the manufacture of proteins by the cell.	Only in eukaryotes.
Smooth endoplasmic reticulum (SER)	A variety of functions, but mainly concerned with the manufacture and transport of lipids.	The endoplasmic reticulum is a system of membrane channels running through the cytoplasm, connected to both the nuclear and plasma membranes. The RER has ribosomes attached to it. Only in eukaryotes.
Rough endoplasmic reticulum (RER)	The manufacture and transport of proteins for **secretion** out of the cell.	
Mitochondrion (pl: mitochondria)	Carries out aerobic respiration.	Only in eukaryotes.
Vesicles	Transport substances to the cell membrane for secretion.	A vesicle is a membrane 'bag'. Only in eukaryotes.
Lysosome	A special type of vesicle which contains protein-digesting enzymes, which can digest its own or invading cells.	Found in old cells near death and white blood cells (to kill bacteria). Only in eukaryotes.
Golgi apparatus	Combines and 'packages' chemicals from the SER and RER.	A stack of membrane sacs. Only in eukaryotes.
Chloroplast	Carries out photosynthesis.	Only in eukaryotes, and only ever found in plant cells.
Vacuole	Storage of sugars.	Large central membrane-bound space in the centre of plant cells.
Cell wall	Provides support.	Made of cellulose in plant cells, peptidoglycan in bacterial cells.
Ribosome	Manufacture of proteins.	Larger in eukaryotic cells, where they are both free in the cytoplasm and attached to the RER.
Flagellum (pl: flagella)	Moves the cell.	More often found in prokaryotic cells, but occasionally in eukaryotic cells.
Nucleoid	The region of prokaryotic cells where most of the DNA is found.	The nucleoid is not a structure, but an 'area'. Cannot really be classed as an organelle.
Plasmid	Controls the manufacture of proteins.	Circular strand of DNA. Only in prokaryotes.
Mesosome	Carries out respiration.	Infolding of a bacterial cell membrane.
Pilus (pl: pili)	Helps some bacterial cells attach to other cells.	
Capsule	Protection of bacterial cells from desiccation and chemicals.	Only in prokaryotes. Polysaccharide layer outside the cell wall. There may also be a 'slime' layer on its surface.

Secretion The release of a substance from the inside of a cell to the outside.

Making links

You can learn more about the characteristic structural features of prokaryotes and eukaryotes in Unit 6, page 158.

Exam tip

You should familiarise yourself with the appearance of these organelles in electron micrographs (photos taken using an electron microscope) as well as in diagrams.

Check your understanding and progress at **www.hoddereducation.co.uk/myrevisionnotes**

1 Which of these features would NOT be found in eukaryotic cells? Mesosome, chloroplast, ribosome, mitochondrion, capsule, endoplasmic reticulum.

2 Both bacteria and plant cells have a cell wall. What features of these structures suggest they may have had a different origin?

3 State one similarity and one difference between ribosomes in prokaryotic and eukaryotic cells.

DNA Deoxyribonucleic acid; a self-replicating nucleic acid which is found in nearly all living organisms and provides a chemical code for the formation of proteins.

Enzyme A protein molecule which catalyses a chemical reaction in the body.

RNA Ribonucleic acid; a single-stranded nucleic acid which plays a role in protein synthesis.

Nucleic acids: DNA and RNA are central to life

Deoxyribonucleic acid (DNA) is a chemical which is central to life. It forms a chemical code which instructs ribosomes how to make specific proteins. Many of these proteins are enzymes, which control all the chemical reactions in the body. As DNA is in the nucleus, but the ribosomes are in the cytoplasm, DNA is partially copied as another nucleic acid, ribonucleic acid (RNA), which can pass through the nuclear membrane and into the cytoplasm. You do not need to know the mechanism for the manufacture of proteins, but you do need to be aware of the structures of DNA and RNA.

DNA consists of two chemical chains, held together by nitrogenous bases, twisted into a double helix shape. Its structure is shown in Figure 1.2.

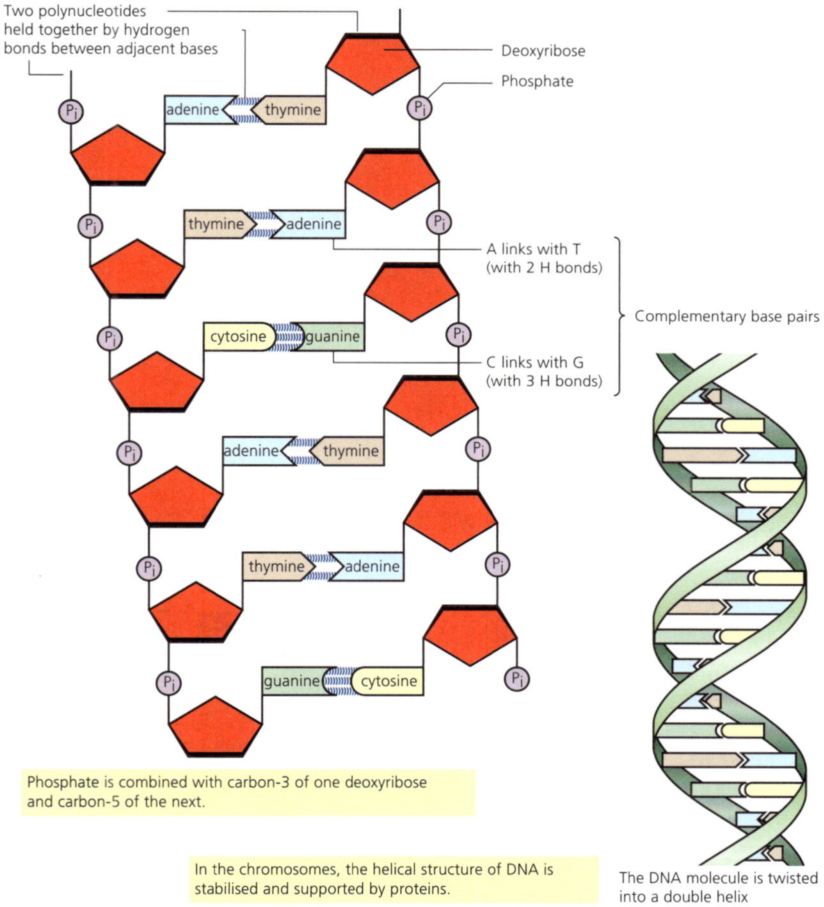

Two polynucleotides held together by hydrogen bonds between adjacent bases

Deoxyribose
Phosphate

adenine — thymine
thymine — adenine

A links with T (with 2 H bonds)

cytosine — guanine

Complementary base pairs

C links with G (with 3 H bonds)

adenine — thymine
thymine — adenine
guanine — cytosine

Phosphate is combined with carbon-3 of one deoxyribose and carbon-5 of the next.

In the chromosomes, the helical structure of DNA is stabilised and supported by proteins.

The DNA molecule is twisted into a double helix

Figure 1.2 The double strand of DNA showing the antiparallel chains

Features of DNA

+ The strands are made of alternating sugar (deoxyribose) and phosphate molecules.
+ The bases which hold the chains together are attached to the deoxyribose.
+ There are four bases – **adenine**, **cytosine**, **guanine** and **thymine**.
+ They form specific pairs – adenine and thymine always pair together, as do cytosine and guanine.

13

- The basic 'unit' of DNA is a **nucleotide**. A nucleotide consists of a phosphate group, a pentose sugar and its attached base.
- DNA is a very long molecule (in humans, between 50 million and 260 million nucleotides in length).

Features of RNA

RNA has the same structure as a single strand of DNA, but with the following differences:

- RNA is not coiled into a helix.
- The sugar is ribose instead of deoxyribose.
- Thymine is not found in RNA. Another base, **uracil**, takes its place.
- The RNA molecule is much shorter (only a few thousand nucleotides in length).

Now test yourself   TESTED ○

4 Name the chemical molecules that make up a nucleotide.

5 State two **chemical** differences between DNA and RNA.

You can calculate magnification as $\dfrac{\text{observed size}}{\text{actual size}}$

If we know the magnification used in a light or electron microscope, it is easy to calculate the actual size of the object using the equation:

$$\text{actual size} = \frac{\text{observed size}}{\text{magnification}}$$

Maths skills

The magnification equation allows us to calculate the magnification of a diagram/photo (if we know the actual size of the structure) or the actual size of a structure (if we know the magnification).

Worked example

An object's image down a light microscope (magnification ×40) is measured as 0.5 mm. Calculate the actual size of the object.

$$\text{actual size} = \frac{\text{observed size}}{\text{magnification}} = \frac{0.5}{40} = 0.0125 \text{ mm}$$

The same equation can be rearranged to calculate the magnification if we know the actual size of an object (this is done to calculate the magnification of a drawing or electron micrograph, as when using a microscope, we know the magnification).

Worked example

Red blood cells have an average diameter of around 7.5 μm (0.0075 mm). n a photograph of red blood cells down a microscope, the average cell diameter is 3 mm. What is the magnification of the photo?

$$\text{magnification} = \frac{\text{observed size}}{\text{actual size}} = \frac{3.0}{0.0075} = \times 400$$

Practice questions

1 A structure in an electron micrograph is 2 cm (20 000 μm) long. The magnification is given as ×3000. What is the real length of the structure?

2 An electron micrograph of a cell is printed in a textbook. The microscope magnification used is ×20 000. Explain why the magnification given for the photograph in the textbook is ×40 000.

Exam tip

Note that the magnification of a drawing or photograph is not necessarily the same as the magnification used by the microscope. Drawings and photographs are rarely 'life size'.

Transport mechanisms

The cell membrane surrounds the cell

All living cells have a cell (plasma) membrane. In eukaryotic cells it consists of a lipid bilayer with protein molecules embedded in it. The structure (known as the fluid mosaic model) is shown in Figure 1.3.

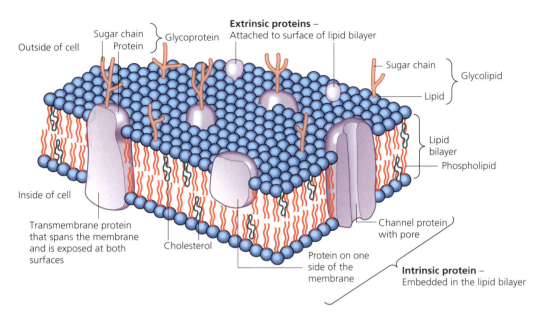

Figure 1.3 Structure of a eukaryotic plasma membrane

Most of the membrane consists of a double layer of phospholipids, known as a *bilayer*. Phospholipid molecules have a *hydrophilic head* which readily mixes with water, and a *hydrophobic tail* which is repelled by water. They naturally form a bilayer in water as it is the most stable structure. The heads are towards the outside in the aqueous medium and the tails are as far away from it as possible; see Figure 1.4). Any substance which is soluble in lipids can easily get through this bilayer, but water-soluble substances cannot.

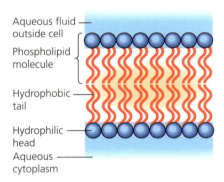

Figure 1.4 Structure of the phospholipid bilayer

Aqueous medium A liquid which contains water, i.e. water or an aqueous solution.

Concentration gradient The difference in the concentration of a solute between two areas. The bigger the difference, the 'steeper' the concentration gradient.

Intrinsic protein A protein embedded in the lipid bilayer of the cell membrane, sometimes completely penetrating it.

Now test yourself

TESTED

6 Which two classes of chemical make up the bulk of the cell membrane?

7 The cell membrane is more permeable to lipid-soluble chemicals than to water-soluble chemicals. Suggest a reason for this.

8 Explain how the hydrophilic heads and hydrophobic tails of lipid molecules lead to the bilayer structure seen in cell membranes.

Unit 1 Key concepts in the application of science

15

Transport by intrinsic proteins allows water-soluble substances to pass through the cell membrane

The presence of intrinsic proteins spanning the width of the membrane allows water-soluble substances to pass through. The intrinsic proteins are of two types, *channel proteins* and *carrier proteins*. The two types transport substances in slightly different ways.

Substances naturally move from an area of higher concentration to an area of lower concentration if there is no barrier in the way, by *diffusion*. Lipid molecules can diffuse through the membrane at any point in the lipid bilayer, but water-soluble (polar) molecules can only diffuse through where there is an intrinsic protein. This conditional form of diffusion is called *facilitated diffusion*. Channel proteins have a pore which allows water-soluble substances to pass through the membrane (Figure 1.5). The channel proteins mainly transport ions and small polar molecules.

Carrier proteins can transport substances by facilitated diffusion (including larger molecules) but can also transport molecules against a concentration gradient (i.e. from a lower concentration to a higher concentration) by *active transport*. Active transport requires energy in the form of *adenosine triphosphate (ATP)* (Figure 1.6).

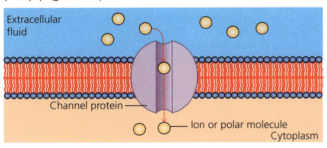 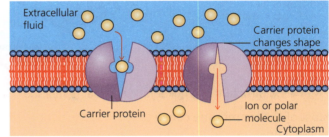

Figure 1.5 Facilitated diffusion

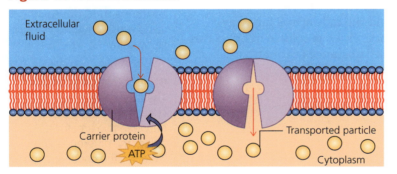

Figure 1.6 Active transport by carrier proteins

Channel proteins can transport any small water-soluble substance, but carrier proteins are specific and only transport a specific substance or group of substances.

Extrinsic proteins have many functions, including acting as antigens and receptors

Extrinsic proteins can be pure proteins or glycoproteins. Glycoproteins are proteins with a short carbohydrate chain attached. Extrinsic proteins have a variety of functions, but the main ones are as follows:

+ Acting as **antigens** (molecules which allow the cell to be recognised by the immune system).
+ Acting as **receptors**. There are many instances where a chemical (e.g. hormones) needs to act on some cells but not on others. Receptors allow the chemical to detect the cells which it needs to affect or not affect.

Check your understanding and progress at **www.hoddereducation.co.uk/myrevisionnotes**

9 Explain the difference between simple diffusion and facilitated diffusion.
10 State **two** differences between the processes of active transport and facilitated diffusion.
11 What is an antigen?

The heart

REVISED ◯

The structure of the heart ensure blood flows through it in one direction

The structure of the mamallian heart is shown in Figure 1.7.

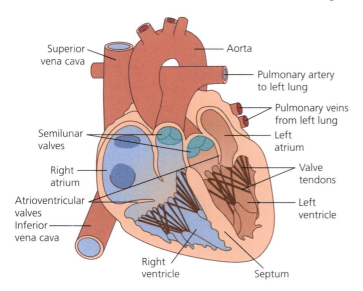

Figure 1.7 Internal structure of the mammalian heart

The valves in the heart ensure that blood always flows through it in the right direction. The *bicuspid* and *tricuspid* valves prevent back-flow from the ventricles into the atria, and the *semilunar* valves stop blood flowing back into the heart from the aorta and pulmonary artery. Some diseases of the heart are caused by faults in these valves, and artificial valves sometimes need to be fitted.

Control of the heart rate is essential

In an adult the average resting heart rate is about 70 beats per minute. This regular rhythm is controlled entirely from within the heart, but sometimes the rate needs to be adjusted. For example, during exercise the muscles have a high demand for oxygen carried in the blood so the heart rate is increased to supply this. This adjustment is carried out using external stimulation by the nervous system.

The cardiac muscle of the heart is unusual in that it can carry out *myogenic contraction*. This is muscle contraction without any external nervous stimulation. The contraction is initiated by the *sinoatrial node* (SAN), a patch of specialised muscle tissue in the upper right atrium (see Figure 1.8).

Exam tip

Always use the term (electrical) impulse when describing how the heart contracts. Vague terms like 'message' or 'signal' are unlikely to be acceptable.

17

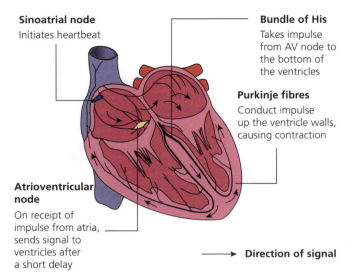

Sinoatrial node
Initiates heartbeat

Bundle of His
Takes impulse
from AV node to
the bottom of
the ventricles

Purkinje fibres
Conduct impulse
up the ventricle walls,
causing contraction

Atrioventricular node
On receipt of
impulse from atria,
sends signal to
ventricles after
a short delay

⟶ **Direction of signal**

Figure 1.8 Co-ordination of the heartbeat

The sinoatrial node (SAN) generates electrical impulses (*myogenic stimulation*) which spread through the atria, causing them to contract. Between the atria and the ventricles there is non-conductive tissue. This stops the spread of the impulse to the ventricles, except at the *atrioventricular node*. After a short delay to allow the atria to complete their contraction, the atrioventricular node sends an impulse to the ventricles. This impulse travels down specialised muscle fibres known as the *bundle of His* to the bottom of the ventricles and then back up through another set of muscle fibres known as the *Purkinje fibres*. As the impulse travels up, it causes the ventricles to contract, forcing blood up into the aorta and pulmonary artery.1

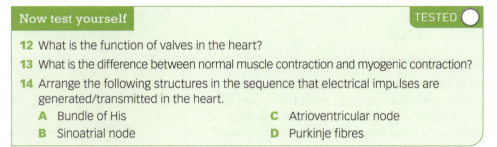

Now test yourself TESTED ⬤

12 What is the function of valves in the heart?

13 What is the difference between normal muscle contraction and myogenic contraction?

14 Arrange the following structures in the sequence that electrical impulses are generated/transmitted in the heart.

  A Bundle of His **C** Atrioventricular node
  B Sinoatrial node **D** Purkinje fibres

The SAN acts as the heart's pacemaker and keeps a steady rhythm. It is not capable, however, of adjusting that rhythm. If the heart rate needs to change (e.g. during exercise) the heart requires external stimulation by the *autonomic nervous system*; for this to happen, specialised sense organs in the body must detect that the rate needs to change.

Chemoreceptors in the heart and in nearby large arteries can detect the levels of carbon dioxide in the blood. A high level of carbon dioxide results from increased respiration and is therefore linked with a low level of oxygen. The heart rate needs to increase to supply the active tissues with enough oxygen. High carbon dioxide levels are detected by the chemoreceptors which send nerve impulses to the cardiac area of the brain. This results in stimulation by the *sympathetic nervous system* which increases the heart rate. The sympathetic nervous system is part of the autonomic nervous system, which controls involuntary functions. The other part, the *parasympathetic system*, causes opposite effects to the sympathetic system.

Blood pressure is sensed by baroreceptors in the aorta and the carotid arteries which detect how much the artery wall is being stretched (higher blood pressure results in more stretch). Impulses are sent to the brain, which will increase the rate and strength of the heart beat if the blood pressure is too low (using the sympathetic nervous system) or reduce the rate and strength if it is too high (using the parasympathetic nervous system).

Chemoreceptor Sense organ which detects a chemical.

Baroreceptor Sense organ which detects pressure.

Check your understanding and progress at **www.hoddereducation.co.uk/myrevisionnotes**

Artificial pacemakers can correct irregular heartbeats

Sometimes a fault in the sinoatrial node or associated structures leads to an irregular heartbeat, a condition known as *arrhythmia*. The patient must be fitted with an electronic artificial pacemaker to correct the condition. Pacemakers are made up of a long-lasting battery and a tiny computer (a *pulse generator*) in a metal case that is fitted underneath the skin. The pacemaker is connected to the heart muscle by leads which can detect heart contractions and send electrical impulses to the muscle. If the heart rate drops below the desired rate, the pacemaker generates an electrical impulse that causes the heart muscle to contract in the correct rhythm.

There are three types of pacemaker:
1 *Single chamber* pacemakers have one lead going to either the right atrium or the right ventricle.
2 *Dual chamber* pacemakers have two leads, one each going to the right atrium and the right ventricle.
3 *Biventricular* pacemakers have three leads, one each going to the right atrium, the right ventricle, and the left ventricle.

The type of pacemaker fitted depends on the specific heart problem the patient has. As a rule, the more serious the condition, the more leads are used. More leads allow for greater control and synchronisation of the heartbeat, but the operation to fit them becomes more complicated.

Leadless micropacemakers are now being developed. These are small devices fixed to the heart which can be fitted by a much simpler (and therefore safer) surgical procedure. However, there is some risk of damaging the wall of the heart and complications can occur if the pacemaker becomes dislodged, which is more likely than with the older types. The first micropacemaker was developed as recently as 2014 and research and development are ongoing.

> **Making links**
>
> You can learn more about the role of science in medicine in Unit 3 The role of biomedical scientists.

> **Now test yourself** TESTED ○
>
> 15 Which part of the autonomic nervous system causes an increase in the heart rate?
>
> 16 Which structure in the heart is likely to be faulty if a single chamber pacemaker is fitted?

Homeostasis REVISED ●

Homeostatic control mechanisms control such things as body temperature and blood pH

The body has physiological control systems that maintain the internal environment within restricted limits. Examples of this include:
+ body temperature range (35.8–37.5 °C)
+ blood glucose range (82–110 mg/dL)
+ blood pH range (7.35–7.45).

To achieve homeostasis, the body has means to detect the current values, and mechanisms to adjust them if they go outside the safe range. These mechanisms often involve hormones (see below).

Many homeostatic control systems are carried out by a mechanism known as negative feedback. An example is the control of water retention (*osmoregulation*) in the body involving anti-diuretic hormone (ADH), the kidney, and the hypothalamus and pituitary gland in the brain. The kidney controls the water content of the blood by reabsorbing more or less water from the urine. The water is reabsorbed as the urine passes through a structure called the collecting duct, and the permeability of the collecting duct wall can be altered by ADH. The mechanism is shown in Figure 1.9.

> **Homeostasis** The maintenance of a constant internal state within the body.
>
> **Hypothalamus** An area in the floor of the brain that maintains the body's internal balance, often by stimulating the release of hormones from the pituitary gland.
>
> **Negative feedback** A process where a change causes a series of events which reverse that change.
>
> **Pituitary gland** A gland hanging from the floor of the brain which produces hormones which control the activity of endocrine (hormone-producing) glands around the body.

19

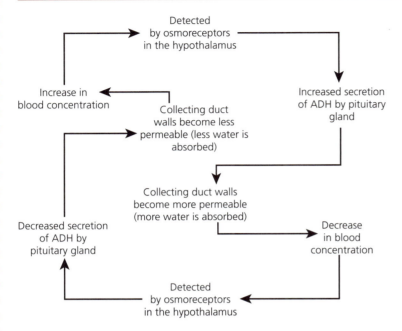

Figure 1.9 Negative feedback for release of ADH

Hormones control many body functions

Hormones are chemicals produced by specialised glands (endocrine glands) which have an effect elsewhere in the body. They travel around the body in the blood. Hormones are important in homeostasis, and also in other aspects of body function. We have already seen how ADH controls water retention in the body. Some other examples of hormones and their functions are given below.

Insulin is produced in the pancreas and lowers blood sugar levels if they get too high. Excess glucose is stored as *glycogen* in the liver and, when needed, the glycogen is broken down to glucose once again. Insulin lowers blood sugar by activating the enzymes which convert glucose into glycogen.

Glucagon is also produced in the pancreas and has the opposite effect to insulin, raising blood sugar levels if they fall too low. It activates the enzymes which convert glycogen to glucose in the liver.

Adrenaline also affects blood sugar levels. It activates enzymes in the same way as glucagon to boost blood glucose levels in stressful situations and during exercise. It is not involved in the routine maintenance of blood glucose levels. Adrenaline is produced in the inner region of the *adrenal glands* – the *medulla*.

Aldosterone is a hormone produced in the *cortex* (the outer region) of the adrenal glands. (These are paired endocrine glands located on top of the kidneys. Each consists of two regions, an inner cortex and an outer medulla.) Aldosterone helps to regulate blood pressure mainly by increasing the amount of sodium reabsorbed into the bloodstream by the kidney and the colon and boosting the amount of potassium in the urine. As water is reabsorbed along with the sodium, this increases blood volume and causes an increase in blood pressure.

The treatment of diabetes differs for type 1 and type 2 diabetes

There are two types of diabetes, known as type 1 and type 2. Both result in an inability of the body to control blood sugar levels, but the causes and treatments of the two types are different.

Type 1 diabetes:

+ Causes are uncertain. Type 1 diabetes is an autoimmune disease, which means that the body's immune system attacks its own cells, in this case specialised cells called beta cells in the Islets of Langerhans in the pancreas, which produce insulin.
+ Patients with type 1 diabetes produce little or no insulin.
+ Onset of type 1 diabetes is usually in childhood or young adulthood.
+ Type 1 diabetes is treated by regular insulin injections and control of carbohydrate intake.

Type 2 diabetes:

+ Causes are uncertain but are known to be linked with obesity and lack of physical activity.
+ Some patients with type 2 diabetes may have reduced levels of insulin, but the main problem is that the body has become insensitive to insulin, so it does not work effectively.
+ Onset of type 2 diabetes is usually in middle age, although recently there has been a rise in younger people with the condition.
+ Type 2 diabetes is usually treated by a strict diet to restrict carbohydrate intake and by tablets, although insulin injections may also be used in some cases.

The key clinical symptom of diabetes is sugar in the urine. This can quickly be detected by the use of urine dipsticks, which change colour according to the level of sugar in the urine and are then compared with a standard colour chart. The diagnosis can be confirmed and refined by use of blood glucose 'pinprick' tests, which provide a digital reading of blood glucose levels. In diagnosis, *fasting* glucose levels are used, where the patient eats nothing for 8–12 hours before the test. Eating carbohydrate influences blood glucose levels and a fasting glucose test eliminates that variable.

Urine dip sticks used to be used by patients to monitor their glucose levels but these have largely been replaced by blood pinprick tests which give a direct reading of blood glucose.

The kidney filters the blood to excrete waste and control salt levels

The kidney has two functions, excretion of wastes and osmoregulation. Osmoregulation involves both the control of water content of the blood and the salt concentration.

The kidney consists of millions of microscopic tubules called *nephrons*.

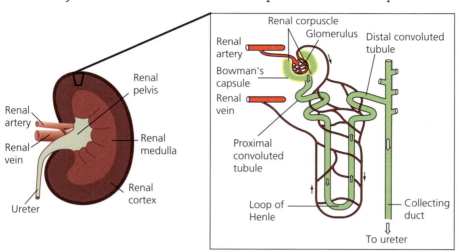

Figure 1.10 Nephron structure and location in the kidney

The kidney functions by filtering the blood (a process known as ultrafiltration) and then reabsorbing useful small molecules that pass through the filter. Essential molecules like glucose are all reabsorbed in the proximal convoluted tubule, but salts such as sodium are only partly reabsorbed, just enough to maintain a suitable level in the blood.

> **Ultrafiltration** Filtration of small molecules.

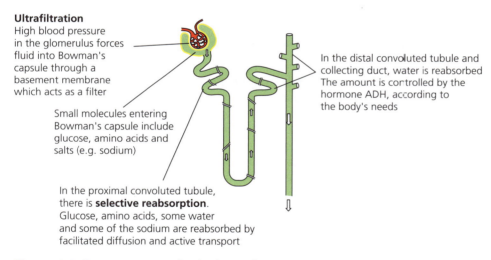

Ultrafiltration
High blood pressure in the glomerulus forces fluid into Bowman's capsule through a basement membrane which acts as a filter

Small molecules entering Bowman's capsule include glucose, amino acids and salts (e.g. sodium)

In the proximal convoluted tubule, there is **selective reabsorption**. Glucose, amino acids, some water and some of the sodium are reabsorbed by facilitated diffusion and active transport

In the distal convoluted tubule and collecting duct, water is reabsorbed The amount is controlled by the hormone ADH, according to the body's needs

Figure 1.11 Processes occurring in the nephron

As outlined earlier in this chapter, the amount of sodium absorbed in the kidneys is controlled by the hormone aldosterone, which is produced in the adrenal cortex. Increased aldosterone secretion increases the amount of sodium absorbed.

The control of salt is important to health

If the concentration of salt in the blood falls below or rises above the normal range, health problems result.

Sodium chloride (salt) deficiency can result from drinking too much water, chronic and severe vomiting or diarrhoea, problems with the heart, liver or kidneys, a deficiency of aldosterone and as a side-effect of certain medications. It is referred to as *hyponatremia* and can result in the following:

+ nausea and vomiting
+ headache
+ confusion
+ loss of energy, drowsiness and fatigue
+ spasms or cramps.

In severe cases, if there is a continuous period of salt deficiency, rapid brain swelling can cause coma and death.

Excess salt in the blood is most likely to result from taking in too much salt in the diet. The main effect is an increase in blood pressure, which increases the risk of cardiovascular disease. The increased concentration of salt in the blood causes water to enter by osmosis, increasing the blood pressure.

> **Now test yourself** TESTED ◯
>
> 22 State the two functions of the kidney.
> 23 Explain why glucose passes from the blood into the nephron, but proteins do not.
> 24 State the part of the body that produces the hormone which controls sodium levels in the blood.

Breathing and cellular respiration REVISED ◯

Breathing and respiration are not the same

In everyday language, breathing and respiration are sometimes used interchangeably, but to a biologist they are distinctly different processes. *Breathing* is a physical, external process which draws air containing oxygen into the body where it is exchanged for carbon dioxide, whereas *respiration* is a chemical, internal process which extracts energy from food in a usable form.

A number of methods are used by scientists to monitor the respiratory system. *Breathing rate* can be easily counted. An instrument known as a spirometer can measure *tidal volume* (the volume of air breathed in and out during normal breathing) and *vital capacity* (the maximum amount of air that can be breathed in and out during deep breathing). A peak flow meter measures how fast a person can breathe out after taking a full breath in.

These tests are used to diagnose and/or monitor respiratory conditions and diseases, including asthma.

Exam tip

In answers, make sure you do not confuse breathing with respiration. Note, however, that the system involved with breathing is called the respiratory system, not the breathing system!

Cellular respiration breaks down food

In the process of cellular respiration, food materials (mainly glucose) are gradually broken down in multiple steps, and the energy within them is used to form adenosine triphosphate (ATP) from adenosine diphosphate (ADP). This addition of phosphate is called a *phosphorylation* reaction. Every cellular process which requires energy uses ATP as its source.

Cellular respiration can be broken down into three stages: glycolysis, which occurs in the cytoplasm; the Krebs cycle, which occurs in the mitochondria; and the electron transfer chain, which also occurs in the mitochondria.

Making links

You learned about mitochondria earlier in this unit. See pages 11 and 12.

The whole process of cellular respiration is shown in Figure 1.12.

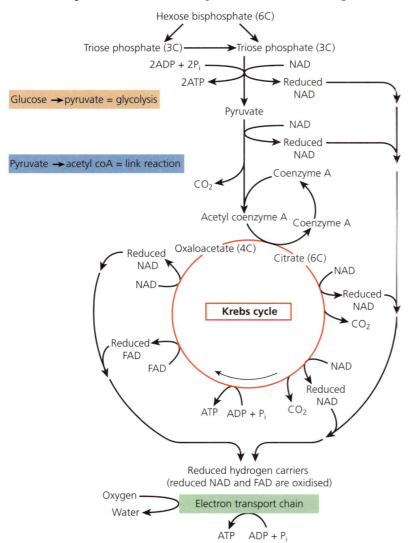

Glucose → pyruvate = glycolysis

Pyruvate → acetyl coA = link reaction

Figure 1.12 An overview of all stages of respiration

Now test yourself   TESTED ⬤

25 Explain the difference between vital capacity and tidal volume.

26 Which stage of cellular respiration does NOT take place in the mitochondria?

27 What type of chemical reaction is involved in the conversion of ADP to ATP?

Key facts about glycolysis

+ Glycolysis takes place in the cytoplasm.
+ The purpose of glycolysis is to convert glucose into pyruvate, which can enter the Krebs cycle.
+ Each of the first two steps (the conversion of glucose to glucose phosphate and then to hexose biphosphate) uses a molecule of ATP.
+ In glycolysis four molecules of ATP are produced, a net gain of two.
+ The production of ATP in this stage is by substrate-linked phosphorylation.
+ Two molecules of NAD are converted into reduced NAD during glycolysis.
+ If oxygen is present, the reduced NAD is fed into the electron transfer chain.
+ The conversion of triose phosphate into pyruvate is an oxidation reaction.

NAD is an abbreviation of nicotinamide adenine dinucleotide. Reduced NAD is sometimes abbreviated to NADH.

Check your understanding and progress at **www.hoddereducation.co.uk/myrevisionnotes**

Key facts about the Krebs cycle

+ The pyruvate formed in glycolysis is converted to acetyl coenzyme a, which enters the Krebs cycle. This reaction is known as the link reaction.
+ Acetyl CoA reacts with a 4-carbon molecule to form a 6-carbon molecule.
+ A series of oxidation–reduction reactions release hydrogen, which will provide electrons for the electron transfer chain.
+ The hydrogen attaches to NAD (see above) or flavine adenine dinucleotide (FAD). NAD and FAD belong to a group of chemicals called *coenzymes*.
+ For each turn of the Krebs cycle, one molecule of ATP is formed by substrate-linked phosphorylation (so two in total for every glucose molecule).
+ Carbon dioxide is released during the Krebs cycle.

Key facts about the electron transfer chain

+ Electrons are passed along a series of electron carriers in the inner membrane of the mitochondrion.
+ The electrons enter the chain when hydrogen, bought to the membrane by NAD or FAD, is released and broken down into electrons and hydrogen ions (H^+).
+ As the electrons pass from carrier to carrier, energy is released and used to phosphorylate ADP to ATP. This is *oxidative phosphorylation*.
+ The final electron acceptor which takes the electrons out of the chain is oxygen, which combines with the electrons and hydrogen ions to form water.
+ For each reduced NAD molecule brought to the electron transfer chain, three ATP molecules are formed, and two for each reduced FAD.

Now test yourself   TESTED

28 Name the end product of glycolysis.

29 How is the hydrogen generated in the Krebs cycle used in the electron transfer chain?

30 What is the precise location of the electron transfer chain in the mitochondrion?

Anaerobic respiration is used when oxygen is unavailable

If oxygen is unavailable or in short supply, certain cells (e.g. muscle cells) can switch to *anaerobic* respiration. In the anaerobic pathway, the Krebs cycle and electron transfer chain do not function, but glycolysis continues. The pyruvate formed is converted into lactic acid in animals and into ethanol and carbon dioxide in plant cells. No ATP is formed in this stage, and so the total net production from a molecule of glucose is two ATP molecules. In aerobic respiration, approximately 30 molecules of ATP are produced per glucose molecule. If anaerobic respiration continues for some time, lactic acid build-up can result in muscle pain, cramps and muscular fatigue.

Basal metabolic rate is the rate at which the body uses energy

Basal metabolic rate (BMR) is the rate at which the body uses energy at rest to maintain basic life functions. It can be measured in kcal/day or kJ/day and varies among individuals. Its measurement is useful in weight control programmes, to help determine a suitable daily calorie intake for the patient.

25

BMR can be determined by direct or indirect methods. BMR may be measured by gas analysis through direct calorimetry. This is the measurement of the heat production of an individual, when placed in an insulated chamber where the heat is transferred to surrounding water. This is the most accurate method of measuring BMR. Indirect calorimetry can also be used. This calculates heat that living organisms produce by measuring either their production of carbon dioxide and nitrogen waste or their consumption of oxygen. Heart rate at rest can be used to estimate energy expenditure because there is a correlation between heart rate and oxygen consumption.

BMR measurements vary among individuals but BMR is generally higher in males than in females, and tends to decrease with age. It is possible that obesity might increase BMR.

Now test yourself   TESTED ⦾

31 Suggest a reason why the electron transfer chain cannot function without oxygen.

32 Suggest why the measurement of carbon dioxide production to establish BMR is referred to as *indirect* calorimetry.

Photosynthesis and food chain productivity   REVISED ⦾

The process of photosynthesis uses carbon dioxide and water to make glucose and oxygen

Photosynthesis involves a complex series of chemical reactions (see below), but can be summarised by the following equation:

carbon dioxide + water → glucose + oxygen

Carbon dioxide and water are the essential raw materials need for photosynthesis, along with light as an energy source. Carbon dioxide is obtained from the air (although the carbon dioxide produced by the plant in respiration can also be used) and water comes from the soil. Photosynthesis takes place in the chloroplasts in leaves. Carbon dioxide can enter the leaves directly through pores called stomata, and water enters the plant via the roots and is carried up the stem to the leaves.

The biochemistry of photosynthesis has a light-dependent stage and a light-independent stage

The biochemistry of photosynthesis can be broken down into two stages:

1 The *light-dependent stage*, in which light energy is used to hydrolyse water into hydrogen and oxygen. The hydrogen provides electrons which are essential for the manufacture of ATP, and eventually joins with nicotinamide adenine dinucleotide phosphate (NADP) to form reduced NADP. The oxygen is released as a waste product.
   In the light-dependent stage, ATP is formed in an electron transfer chain similar to that in respiration.

2 The *light-independent stage*, in which the ATP and reduced NADP formed in the light-dependent stage are used to produce carbohydrate (mainly glucose). This process also requires carbon dioxide.
   The glucose formed in the light-independent stage can be processed in the plant to produce the other essential foodstuffs, proteins (with the addition of nitrogen) and lipids.

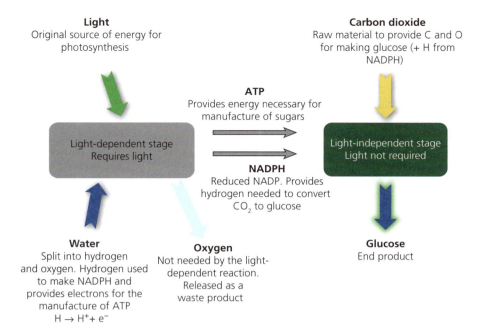

Light
Original source of energy for photosynthesis

Carbon dioxide
Raw material to provide C and O for making glucose (+ H from NADPH)

ATP
Provides energy necessary for manufacture of sugars

Light-dependent stage
Requires light

Light-independent stage
Light not required

NADPH
Reduced NADP. Provides hydrogen needed to convert CO_2 to glucose

Water
Split into hydrogen and oxygen. Hydrogen used to make NADPH and provides electrons for the manufacture of ATP
$H \rightarrow H^+ + e^-$

Oxygen
Not needed by the light-dependent reaction. Released as a waste product

Glucose
End product

Figure 1.13 Summary of photosynthesis

> **Now test yourself** TESTED ◯
>
> 33 Name the three products of the light-dependent stage of photosynthesis.
> 34 Name the organelle that carries out photosynthesis.
> 35 The light-independent stage does not directly require light, yet it cannot continue for long in the dark. Suggest an explanation for this.
> 36 Which mineral is required for the plant to make proteins from the products of photosynthesis?

Photosynthesis is fundamental to food chains

Within an ecosystem, the organisms are inter-linked by a series of food chains, forming a food web.

Food chains consist of stages known as trophic levels (usually 4 or 5). These are shown below:

> **Food chain** The sequence of transfers of energy (in the form of food) from organism to organism.
>
> **Food web** The interconnection of food chains in an ecosystem.
>
> **Trophic level** The organisms in an ecosystem which occupy the same level in a food chain.

Trophic level →	1st consumer →	2nd comsumer →	3rd consumer →	4th consumer →	5th consumer
Description	Organism which produces food (green plant)	Feeds on the producer (herbivore)	Feeds on the 1st consumer (carnivore)	Feeds on the 2nd consumer (carnivore)	Feeds on the 3rd consumer (carnivore)

As green plants are the organisms which make food by photosynthesis, they are always producers and initiate every food chain. Carnivores can occupy different trophic levels at different times, depending on what type of animal they eat.

The efficiency of food chains decreases along the chain

Energy is transferred along food chains, but the transfer can never be 100% efficient and some energy is lost along the way. Humans function as consumers (usually 1st or 2nd, but sometimes 3rd or 4th) and so it is in our interest to ensure that the energy transfer is as efficient as possible.

> **Exam tip**
>
> If a question asks what the arrows in a food chain show, it is the flow of energy. 'It shows what eats what' is not acceptable.

Constraints on the efficiency of photosynthesis
+ Some sunlight does not hit the leaves of plants, and so is not absorbed.
+ Some sunlight is reflected from the leaves, so is not absorbed.

27

- Some sunlight travels right through the leaves (transmission) and is not absorbed by chloroplasts.
- Cold temperatures slow down the chemical reactions in photosynthesis.
- Photosynthesis needs carbon dioxide and water. If there is a shortage of either, photosynthesis will be restricted.
- In crowded conditions, plants may shade others, and the shaded ones cannot photosynthesise at their maximum rate.
- Plants need adequate nutrients (minerals) to remain healthy, and specifically nitrogen if they are to make proteins and grow.

The total amount of energy converted into food by plants is referred to as *gross primary production* (GPP). Not all of this energy is passed on to primary consumers because the plants will use much of it for their own purposes (e.g. growth). The amount that is available to be passed on is the *net primary production* (NPP).

Constraints on the efficiency of food chains

Every time energy is transferred from one trophic level to the next, some energy is lost.

Every organism uses some of the energy it takes in for its own purposes (by means of respiration), e.g. for growth, excretion and (in animals) movement.

- The feeding organism often does not eat the whole organism. Animals usually feed on plant shoots or roots, but not both. Larger animals are rarely eaten whole.
- The feeding organism is often incapable of digesting all the organism it eats. Cellulose in plants is difficult to digest. The skeletons of small animals and the hard parts of insects are rarely digested.

In general, only about 10% of the energy taken in by an organism is passed on to the next trophic level.

Now test yourself   TESTED

37 Explain why the rate of photosynthesis in a plant is lower in winter.

38 Why is net primary production always lower than gross primary production?

39 Why is plant food rarely digested completely by animals?

40 Assume that 90% of the energy in a trophic level is lost before transfer to the next trophic level. The producers in an area contain 21 564 kJ of energy. How many kJ will eventually reach the 2nd consumers?

Ecological pyramids show how productive different trophic levels are

The productivity of different trophic levels can be shown by ecological pyramids. An example of an *energy pyramid* is shown below. The width of each block indicates the total amount of energy in each trophic level.

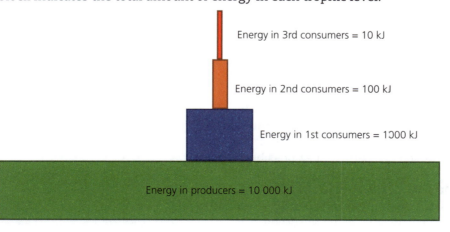

Energy in 3rd consumers = 10 kJ

Energy in 2nd consumers = 100 kJ

Energy in 1st consumers = 1000 kJ

Energy in producers = 10 000 kJ

Figure 1.14 Example pyramid of energy

Check your understanding and progress at **www.hoddereducation.co.uk/myrevisionnotes**

There are also *pyramids of biomass* indicating the dry mass (the mass of an organism or group of organisms not including their water content) of each trophic level.

Pyramids of energy are the most accurate as they measure energy directly but require the bodies of the organisms to be incinerated and the energy given off measured using a calorimeter.

Pyramids of biomass require measurement of dry mass by heating the organism(s) in an oven until they reach constant weight. This obviously kills the organisms.

Meat-free diets have advantages and disadvantages

Advantages of a meat-free diet

+ As each link in a food chain results in a loss of energy, it is more energy-efficient to eat organisms lower in the food chain. If we eat plants, we obtain approximately 10% of their energy. If we eat animals that have eaten plants (e.g. beef, pork, lamb and chicken) we receive only 1% of the energy that was originally in the plants.
+ Most meats are high in saturated fats, which raise cholesterol levels, causing an increased risk of heart disease.
+ In general, meat-free diets are lower in calories and are less likely to lead to obesity (which is linked to many health conditions).
+ Farm animals are often given antibiotics and growth hormones, which may remain in the meat when eaten.
+ Farm animals (especially cows) release a lot of methane, a greenhouse gas which contributes to global warming.

Disadvantages of a meat-free diet

+ Meat contains all the amino acids we need for good health. No single plant does, so the amino acids can only be obtained from a vegetarian diet if it contains a significant variety of different foods.
+ Depending on the choice of fruits and vegetables, the diet may be deficient in protein.
+ If meat in the diet is replaced by ready meals, processed food and a lot of cheese and dairy products, the meat-free diet may be less healthy.

Now test yourself   TESTED ◯

41 Suggest a reason why dry mass is used when constructing pyramids of biomass.

42 Why is it more energy efficient to eat plants rather than animals?

43 Explain why a vegetarian who eats a very restricted range of fruits and vegetables might damage their health.

Summary

+ There are two types of cell – prokaryotic and eukaryotic.
+ Prokaryotic and eukaryotic cells have a cytoplasm and a cell membrane but other organelles are unique to one type or the other.
+ All organelles found in cells have a specific function.
+ Structures you need to know are: nuclei, smooth endoplasmic reticulum (SER), rough endoplasmic reticulum (RER), mitochondria, vesicles, lysosomes, Golgi apparatus, chloroplasts, vacuoles, cell walls, ribosomes (70S and 80S), flagella, nucleoid, plasmids, mesosomes, pili, slime capsules.
+ Cells contain two types of nucleic acid: DNA and RNA.

+ $\text{magnification} = \dfrac{\text{observed size}}{\text{actual size}}$
+ Cell membranes consist of a phospholipid bilayer with proteins interspersed.
+ The proteins in the membrane have roles in transport and cell recognition.
+ Key structures of the heart include the bicuspid valve, tricuspid valve, semilunar valves, sinoatrial node (SAN), atrioventricular node (AVN), Purkinje fibres and bundle of His.
+ The SAN, AVN, Purkinje fibres and bundle of His are used in myogenic stimulation of the heart.

- Carbon dioxide chemoreceptors and baroreceptors have a role in controlling heart rate.
- Artificial pacemakers are used as treatment for arrhythmia (abnormal heart rate), and each type has advantages and disadvantages.
- Homeostasis involves the physiological control systems that maintain the internal environment of the body. There are systems for body temperature, blood glucose concentration and blood pH.
- Negative feedback is a homeostatic mechanism.
- The hormones insulin, glucagon, ADH and aldosterone play a role in body function.
- The pancreas and liver help to regulate blood glucose concentration.
- The body's normal system for regulating blood glucose concentration involves insulin, glucagon and aldosterone.
- Type 1 and 2 diabetes have different causes and different treatments.

- Health professionals and patients with diabetes use physiological measurements to inform diagnosis and treatment of diabetes.
- The hypothalamus, pituitary and ADH play a role in osmoregulation.
- In the nephron within the kidney, the Bowman's capsule acts as an ultrafiltration unit and the convoluted tubules carry out selective reabsorption of glucose, sodium ions and water.
- The adrenal cortex and aldosterone are also involved in the reabsorption of sodium ions.
- Sodium chloride (salt) deficiency has both short term and long-term effects on health.
- A number of circumstances can create a risk of losing too much salt.
- Excess salt in the diet might create health problems.
- Excess or deficiency of ions and hormones have an effect on health.

Exam practice

1   The development of electron microscopes has allowed scientists to study the organelles in living cells. One type of organelle is a mitochondrion.
The figure shows the structure of a mitochondrion.

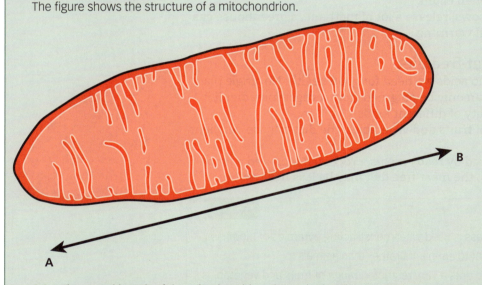

1.1   The actual length of the mitochondrion shown in the diagram is 7 µm. Calculate the magnification of the diagram. [2]

1.2   What is the function of the mitochondria in a cell? [1]

1.3   Mitochondria are found in eukaryotic cells but not in prokaryotic cells. Name one other organelle that is only found in eukaryotic cells. [1]

1.4   Name the structure in prokaryotic cells that has the same function as the mitochondria. [1]

1.5   Mitochondria were first described in 1890 by the German scientist Richard Altmann, but the cristae within them were not described until the 1950s. Suggest a reason for this. [2]

1.6   Some scientists believe that mitochondria were originally a type of bacterium that 'invaded' primitive cells and became established inside them. Mitochondria have their own DNA and 70S ribosomes. Suggest how this evidence supports the theory that they were once bacteria. [3]

2 Patients with an increased risk of heart disease frequently have their blood pressure and pulse rate monitored. The graph shows the readings of systolic blood pressure (when the heart contracts) and pulse rate of one patient over a month.

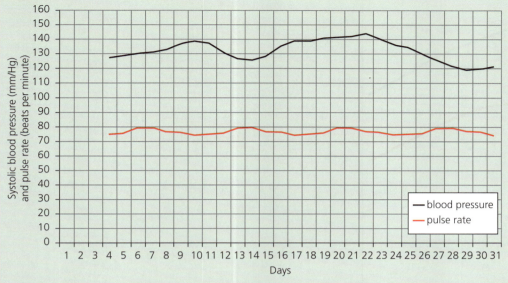

Researchers concluded that a lowering of blood pressure tended to be related to an increase in heart rate.

2.1 State **one** piece of evidence from the graph that supports the researchers' conclusion. [2]

2.2 State **one** piece of evidence from the graph that throws doubt on the researchers' conclusion. [1]

2.3 When taking readings on different days, suggest **two** factors that should be controlled. [2]

2.4 Select the type of sense organ that detects blood pressure from the choices below: [1]

    A baroreceptor   **C** chemoreceptor

    B photoreceptor   **D** osmoreceptor

2.5 The heartbeat is initiated by the sinoatrial node. In which area of the heart is this located? [1]

3 Patients with type 1 diabetes need to monitor the level of sugar in their blood and adjust their insulin doses accordingly. If blood glucose levels get too high, sugar appears in the urine. In the past, patients would test their urine for sugar, but more recently blood testing meters have become available so that the blood can be tested directly. The most recent development in testing is to use an electronic monitor which is attached to the arm, with a sensor that goes below the skin. The sensor tests the level of sugar in the fluid between the cells and produces a reading on a special meter or a mobile phone. When blood glucose levels change, the level of glucose in this fluid will also change a short while later.

3.1 Suggest **two** advantages of testing blood for glucose rather than testing urine. [2]

3.2 Suggest **one** advantage of using an electronic monitor rather than blood testing. [1]

3.3 Low blood sugar levels can make driving dangerous. People with type 1 diabetes are required to test their blood sugar before they drive. Until recently, diabetics were told they must test their blood rather than using their electronic monitor for this purpose. Using the information in the question, suggest a reason for this. [2]

3.4 Type 1 diabetics are treated with insulin injections, but most type 2 diabetics are not. Suggest a reason for this. [2]

3.5 A person with type 1 diabetes works in an office. Every Saturday morning, they go to the local swimming pool and swim for an hour. Before breakfast on a Saturday, they give themselves a slightly lower dose of insulin than on weekdays. Explain the reason for this. [3]

4   A study in 2008 looked at the effect of gender and age on the vital capacity of the respiratory system. The results are shown below.

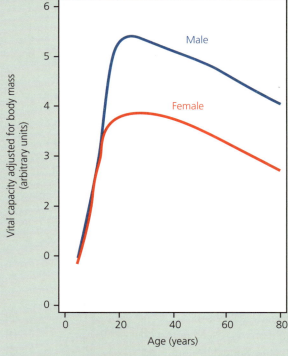

4.1   What is meant by the term vital capacity? [1]

4.2   Name the instrument that would be used to measure vital capacity. [1]

4.3   State **two** conclusions that can be drawn from the results of this experiment. [2]

4.4   Suggest a reason why the results for vital capacity have been adjusted for body mass. [2]

4.5   Oxygen breathed in is used for aerobic respiration. When muscle cells have insufficient oxygen (e.g. during exercise) they can respire anaerobically. Suggest a reason why muscle cells do not respire anaerobically all the time. [2]

4.6   Which stage of respiration occurs whether or not oxygen is present? [1]

5   Plants can be mass produced by a process called tissue culture. Plants are cut into small pieces and each piece is stimulated to grow into a complete plant. The young plants are placed in a room known as a growth room, in which light intensity, carbon dioxide levels, water supply and temperature are controlled throughout the day. Sweet potatoes are an example of a crop that is commonly grown by tissue culture.

5.1   Explain why the process of photosynthesis is essential for plant growth. [2]

5.2   Explain why low temperatures will restrict plant growth. [1]

5.3   Light is required for the light-dependent stage of photosynthesis. State the two products of the light-dependent stage that are required by the light-independent stage. [2]

5.4   Blue and red light are the best colours of light for photosynthesis. Green light is reflected by chloroplasts. Explain why this means that the rate of photosynthesis in green light is extremely low. [2]

5.5   Sweet potatoes are a common component of meat-free diets. Explain why meat-free diets are more energy-efficient for the planet than diets including meat. [4]

5.6   State **one** disadvantage of meat-free diets. [1]

Key concepts in the application of chemistry

Introduction REVISED ⬤

A good knowledge of atomic structure and electron configurations is the basis of understanding chemical structures and reactions. This leads to applications of isotopes, spectrometry and the use of coloured compounds.

Knowing the patterns of the Periodic Table allows you to predict the properties and new applications of elements including the noble gases, semiconductors and transition metals.

Knowledge of the mole and reaction stoichiometries allows you to determine reacting masses and production yields of chemicals and leads to the application of quantitative analysis.

When developing new materials for new contexts, you need an understanding of the structure and bonding of materials.

You also need to know why enthalpy changes and applications of Hess's Law are important in the development of new chemicals and in the analysis of foods. You need to be able to calculate enthalpies of reaction to determine and understand the effect of reaction conditions on yields.

Atomic structure REVISED ⬤

Atoms are the fundamental building blocks of matter

Atoms consist of a tiny nucleus containing protons and neutrons, surrounded by electrons arranged into sub shells. Figure 1.15 shows a diagrammatic representation of a helium atom. This is not drawn to scale: in reality, the nucleus is about $\dfrac{1}{100\ 000}$ of the size of the atom.

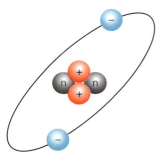

Figure 1.15 A helium atom

Sub-atomic particles are so small that chemists frequently describe their properties relative to each other. Table 1.3 shows the relative masses and charges of each of these particles.

Table 1.3 Relative masses and charges of sub-atomic particles

Sub-atomic particle (symbol)	Relative mass	Relative charge
proton, p	1	+1
neutron, n	1	0
electron, e	$\dfrac{1}{1800}$	−1

Atoms are electronically neutral, so the number of protons in an atom must equal the number of electrons. Atoms that have lost or gained electrons are called ions. Ions have an electrical charge due to the imbalance of protons and electrons.

The number of protons in an atom (or nucleus) is called the *atomic number* (or proton number), Z. The atomic number of helium, shown in Figure 1.15, is 2.

The *mass number*, A, is the number of protons plus the number of neutrons in an atom (or nucleus). The mass number of helium (Figure 1.15) is 4; two protons plus two neutrons.

The $_Z^A X$ notation is used to describe the structure of atoms (and nuclei) where X is the chemical symbol for the element. The $_Z^A X$ notation for helium is $_2^4 He$. Atoms (and nuclei) can also be described with their name and mass number e.g. helium-4 or chlorine-35.

> **Ion** A charged particle formed when an atom loses or gains electrons. Some ions are formed from a small collection of atoms, such as sulfate or carbonate ions.

Now test yourself

TESTED ○

1 Using the Periodic Table, complete this table:

Atom/ion name	$_Z^A X$ notation	Number of protons	Number of neutrons	Number of electrons
Hydrogen-3		1		
Fluorine-19				
Sodium-23⁺ (ion)				
Sulfur-32²⁻ (ion)				

Isotopes are atoms (or ions) of the same element that have different numbers of neutrons but the same number of protons

Helium, for example, has nine known isotopes: $_2^2 He$; $_2^3 He$; $_2^4 He$; $_2^5 He$; $_2^6 He$; $_2^7 He$; $_2^8 H$; $_2^9 He$ and $_2^{10} He$. However, only two of these isotopes, $_2^4 He$ and $_2^3 He$, are stable, and 99.9999% of naturally occurring helium atoms are $_2^4 He$ while only 0.0001% is $_2^3 He$. The other helium isotopes have extremely short half-lives and have only been observed inside nuclear reactors or the core of stars.

The percentage amount of naturally occurring isotopes is called the *isotopic abundance*. Many elements have only one dominant isotope, such as helium-4; however, some elements (such as chlorine) have two isotopes with significant abundances. The two significant isotopes of chlorine are $_{17}^{35} Cl$ and $_{17}^{37} Cl$ and their relative abundances are 76% and 24% respectively. This means that the mean mass number of chlorine is 35.5.

Knowledge of isotopes is very important to certain scientific occupations. Radiographers work in hospitals using radioactive materials to image the body or provide therapeutic treatments. Many isotopes are stable, but some are unstable and can decay by the emission of radiation. Radiographers need to know and understand the chemical properties and decay mechanisms of these unstable isotopes.

> ### Making links
>
> In Unit 6b Medical physics, you will learn about the uses of radioisotopes in medical imaging and therapy.

Now test yourself

TESTED

2 Why are hydrogen-1, deuterium (hydrogen-2) and tritium (hydrogen-3) described as isotopes of hydrogen?

3 Sulfur is known to have four stable isotopes, containing 16, 17, 18 and 20 neutrons. Write out the $_Z^A X$ notation for each of these four isotopes.

Check your understanding and progress at **www.hoddereducation.co.uk/myrevisionnotes**

Electrons are arranged in shells and sub-shells

At GCSE the model for electron arrangement involves shells (or energy levels). Each shell has a maximum occupancy. The GCSE electron structure of sodium is shown in Figure 1.16.

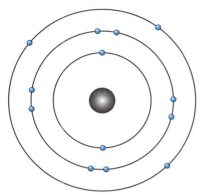

Figure 1.16 GCSE model of electron structure

In this model, sodium atoms have a total of 11 electrons arranged in three energy levels, $n = 1$, $n = 2$ and $n = 3$. The first shell can contain a maximum of 2 electrons, the second shell 8 electrons, the third shell 18 electrons and the fourth shell 32 electrons. This model is a simplification, and a better, more advanced model, further arranges the shells into sub-shells:

+ A sub-shell is an orbital or a combination of orbitals, which describes the places with the highest probability of finding electrons.
+ Each orbital can hold a maximum of two electrons.
+ Although there are four types (or shapes) of orbital, s, p, d and f, only s, p and d are studied in this course.
+ Each energy level (or shell) can contain:
    + one s orbital
    + three p orbitals (the p sub-shell) from the second energy level up
    + five d orbitals (the d sub-shell) from the third energy level up.

A symbolic naming system has been developed to describe the arrangement of electrons in this model. The number of the energy level (1, 2, 3, …) always comes first followed by the orbital letter (s, p, d) and finally, the number of electrons in the particular sub-shell is written as a superscript. As an example, the electron structure of sodium is written as:

$1s^2 2s^2 2p^6 3s^1$

The sub-shells are filled in a particular order, as shown in Figure 1.17.

Figure 1.17 Sub-shells filling order

> **Sub-shell** One of the orbitals that makes up an electron energy shell (energy level).

> **Exam tip**
>
> You need to know the (sub-shell) electron configurations for atoms and ions up to $Z = 36$ (krypton, Kr).

Now test yourself   TESTED ◯

4 What is meant by the term electron sub-shell?

5 What is the maximum number of orbitals permitted at the $n = 2$ energy level?

6 Use the Periodic Table to complete the following table, showing the sub-shell electron configurations of selected atoms:

Atom	Sub-shell electron configuration
Hydrogen	
Oxygen	
Calcium	
Nickel	

My Revision Notes: AQA Applied Science Suitable for Level 3 and Level 3 Extended Certificates

Colour is related to electron structure

Electron configurations can also be represented by energy level diagrams.

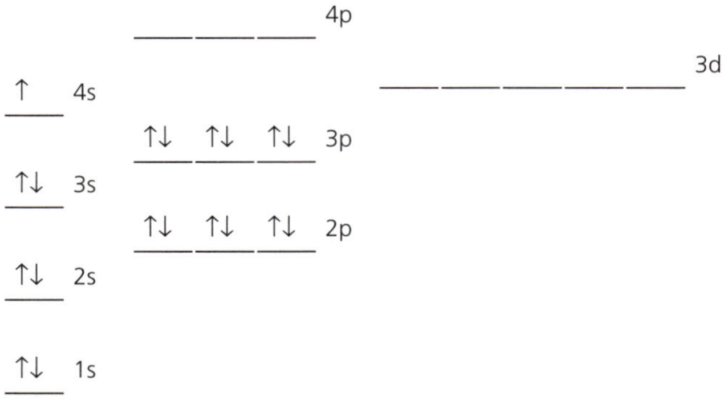

Figure 1.18 The energy level diagram of a potassium atom

In these diagrams, higher energy levels are drawn further up the diagram, and electrons are represented by up ($\uparrow$) and down ($\downarrow$) arrows.

Electrons can move from one sub-orbital to another (empty) sub-orbital. To move to a higher sub-orbital, an electron must gain energy from its surroundings (in the form of a photon of electromagnetic radiation – infrared, visible light or ultraviolet), and to drop down to a lower sub-orbital, the electron must transfer energy out to the surroundings (as photons).

One of the experimental identification techniques used by analytical chemists is a flame test (to produce coloured flame emission spectra). Compounds are heated in a hot flame and electrons within the atoms absorb infrared photons from the flame and move to higher-energy sub-orbitals. The electrons then drop down to lower-energy sub-orbitals, emitting photons of light as they do so. The emitted light is analysed using a spectrometer, and the different atoms can be identified because each atom (or ion) has its own unique electron configuration, giving a unique **emission spectrum**.

> **Emission spectrum** The (unique) pattern of different coloured photons emitted by a hot element.

Some materials made from transition metal compounds are coloured and have uses as dyes, pigments and paints. Compounds appear coloured because they absorb the photons of light of some colours of the visible spectrum, but reflect one particular colour of photon (or a mixture of different coloured photons). The reflected photons give the compound its distinctive colour. The absorbed photons correspond to electron level changes within the atoms.

> **Now test yourself** TESTED ⬤
>
> 7 The electron in the 4s^1 sub-shell of potassium, as shown in Figure 1.81, is excited up to the 4p^1 sub-shell. Explain why three possible photons can be emitted when the electron decays back into the 4s^1 sub-shell.

Masses are calculated using relative masses

The masses of atoms in chemistry are determined relative to the mass of one atom of carbon-12. The mass of one atom of carbon-12 is given as 12.0000.

This means that any 'relative' mass in chemistry is measured against $\frac{1}{12}$ of the mass of an atom of carbon-12.

The *relative atomic mass* (A_r) of an atom of a particular element is its mass relative to $\frac{1}{12}$ of the mass of an atom of carbon-12. These values take into account the masses of the electrons as well as the relative isotopic abundances of the elements, and these are the values displayed on a Periodic Table.

The *relative molecular mass* (M_r) of a molecule is its mass relative to the mass of $\frac{1}{12}$ of the mass of a carbon-12 atom. To calculate the relative molecular mass of a molecule you add up the relative atomic masses of all the atoms in the molecule.

The *relative formula mass* (RFM) of an ionic compound is the sum of all the relative atomic masses of the atoms in the compound.

Worked example

Calculate the relative molecular mass, M_r, of methanol, CH_3OH.

1 Identify all the elements in the molecule and the number of atoms of each:

   C: 1

   H: 4

   O: 1

2 Assign the relative atomic masses to each element and multiply:

   C: $1 \times 12 = 12$

   H: $4 \times 1 = 4$

   O: $1 \times 16 = 16$

3 Add up all of the values:

   Relative molecular mass (M_r) of methanol = $12 + 4 + 16 = 32$

Exam tip

In the examination, it is acceptable to use the term relative molecular mass for ionic compounds.

Worked example

Calculate the RFM of sodium carbonate, Na_2CO_3.

1 Identify all the elements in the molecule and the number of atoms of each:

   Na: 2

   C: 1

   O: 3

2 Assign the relative atomic masses to each element and multiply:

   Na: $2 \times 23 = 46$

   C: $1 \times 12 = 12$

   O: $3 \times 16 = 48$

3 Add up all of the values:

   Relative formula mass (RFM) of sodium carbonate = $46 + 12 + 48 = 106$

Now test yourself   TESTED ◯

8 Why is the relative atomic mass, A_r, of lithium different from the mass number, A, of lithium-7?

9 Use a Periodic Table to calculate the relative molecular mass (M_r) of each of the following molecules:

   a ethene, C_2H_4 c carbon tetrachloride, CCl_4

   b ammonia, NH_3 d propanol, C_3H_7OH

10 Use a Periodic Table to calculate the relative formula masses (RFM) of the following ionic compounds:

   a sodium oxide, Na_2O c aluminium oxide, Al_2O_3

   b potassium sulfate, K_2SO_4 d ammonium hydrogencarbonate, $(NH_4)HCO_3$

The Periodic Table arranges elements in order of increasing atomic (or proton) number

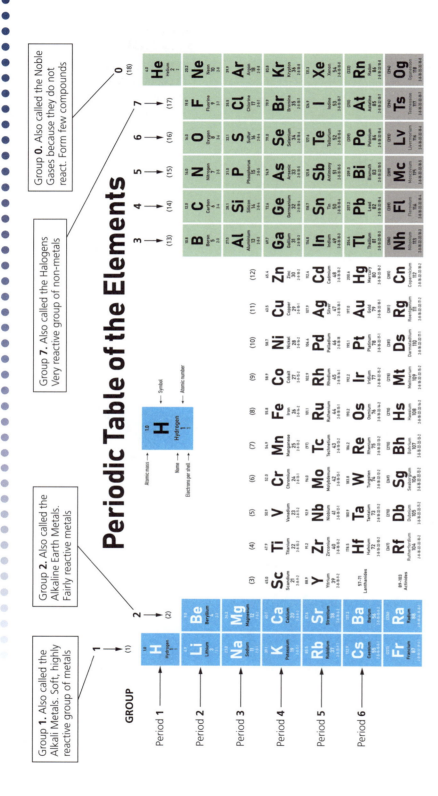

Periodic Table of the Elements

Group 0. Also called the Noble Gases because they do not react. Form few compounds

Group 7. Also called the Halogens Very reactive group of non-metals

Group 2. Also called the Alkaline Earth Metals. Fairly reactive metals

Group 1. Also called the Alkali Metals. Soft, highly reactive group of metals

s-block elements

d-block elements

p-block elements

f-block metals

Figure 1.19 The Periodic Table

Check your understanding and progress at **www.hoddereducation.co.uk/myrevisionnotes**

Rows in the Periodic Table are called *periods*, and elements in the same period contain the same number of electron shells (energy levels). Moving from left to right across the period fills the electron shell up to two (in row/period 1) or eight (subsequent rows/periods), and the elements change from metallic to non-metallic.

Columns in the Periodic Table are called *groups*. Elements in the same group have the same number of outer shell electrons. This means that they all have similar chemical properties.

There are two ways of numbering groups: (a) the International Union of Pure and Applied Chemistry (IUPAC) new (post-1990) system involves numbering the groups from 1 to 17 left to right; and (b) the older IUPAC (pre-1990) system (in which the transition metals are not numbered) involves numbering the other groups from 1 to 7 followed by 0 for the noble gases group. Your specification will generally use the older system, but the newer group numbers are given in brackets, for example, Group 7 (16), the halogens.

Now test yourself TESTED ⭕

11 Explain why the Periodic Table is ordered by atomic number and not atomic mass.

Regions of the Periodic Table

The blocks of the Periodic Table relate to electron configurations and, in particular, the sub-shell in which the outer (highest energy level) electrons are located.

Block An area of the Periodic Table where all the elements have their highest energy level electrons in the same orbital type.

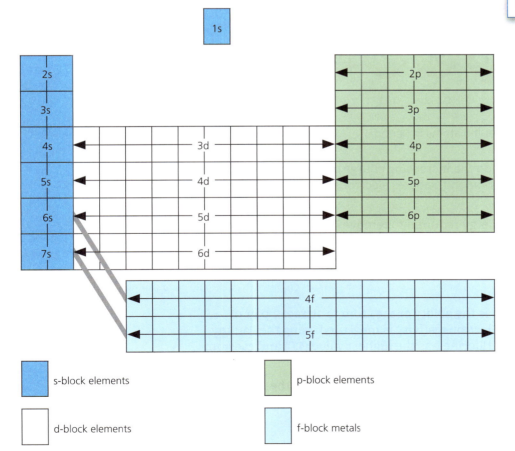

s-block elements

p-block elements

d-block elements

f-block metals

Figure 1.20 The Periodic Table divided into different blocks

My Revision Notes: AQA Applied Science Suitable for Level 3 and Level 3 Extended Certificates

For example, potassium, in Group 1, is an alkali metal. It has the electron configuration: $1s^2\ 2s^2\ 2p^6\ 3s^2\ 3p^6\ 4s^1$: its highest energy level electron is $4s^1$ so it is in the s-block.

Fluorine, in Group 7, is a halogen. Its electron configuration is $1s^2\ 2s^2\ 2p^5$: its highest energy level electron is $2p^5$ so it is in the p-block.

Now test yourself   TESTED ◯

12 Complete the following table.

Group name	Group	Block	Period	Element
Alkali metals			2	
Alkaline earth metals				Ca
Transition metals				Fe
Halogens			3	
	0 (18)		4	

13 Which elements have electrons in the following highest energy levels:
   a $2s^2$ **c** $3d^1\ 4s^2$
   b $3p^1$ **d** $4p^6$?

The properties of the elements change across a period (row)

To illustrate this, Period 3 will be used as an example. This period contains the elements with atomic numbers 11 to 18.

The *atomic radius* decreases across the period, as shown by the scale diagram in Figure 1.21, where the values given are in nanometres (nm) (1 nanometre, nm, is equal to 1×10^{-9} m).

Na	Mg	Al	Si	P	S	Cl	Ar
0.154	0.130	0.118	0.112	0.110	0.102	0.099	0.095

Figure 1.21 Atomic radius (nm) of the atoms in Period 3

This decreasing pattern is due to the increasing charge on the nucleus, which exerts a greater force of attraction on the outer electrons so they are pulled closer to the nucleus.

The ionisation energy is the energy required to remove one mole of electrons from one mole of the gaseous element forming +1 ions. The increase in ionisation energy across the period is due to the increase in the nuclear charge, which decreases the atomic radius and binds the outer electrons to the nucleus with more energy.

The *electronegativity* of an atom is the ability of the atom to attract a pair of electrons in a covalent bond. Electronegativity is measured using the Pauling electronegativity scale, where fluorine has the highest value (4.0) and caesium has the lowest value (0.7).

The electronegativities of the Period 3 elements increase across the period due to the increase in the nuclear charge, which increases the attraction of the outer electrons.

14 The following table shows the Pauling electronegativities of the Period 2 elements, together with their atomic numbers.

Element	Li	Be	B	C	N	O	F
Atomic number	3	4	5	6	7	8	9
Pauling electronegativity value	1.0	1.5	2.0	2.5	3.0	3.5	4.0

   a State the trend in these values.
   b Explain why neon is not included in these values.

> **Exam tip**
>
> Remember, ionisation energy involves losing electrons; electronegativity involves gaining electrons.

Patterns up and down groups of the Periodic Table

s-block metals
The s-block contains two groups:
1 the alkali metals (lithium, Li; sodium, Na; potassium, K; rubidium, Rb; caesium, Cs; and francium, Fr)
2 the alkaline earth metals (beryllium, Be; magnesium, Mg; calcium, Ca; strontium, Sr; barium, Ba; and radium, Ra).

Moving down the groups:
+ the atomic radius increases
+ ionisation energy decreases
+ electronegativity decreases
+ reactivity increases
+ melting and boiling points decrease.

d-block metals
The d-block metals are sometimes referred to as the *transition* metals and are characterised by generally forming coloured compounds. Moving across the d-block, in any given period, the atomic radii decrease and then increase from left to right. Moving down a group, the atomic radii increase. There is an irregular trend of ionisation energies across the d-block. There are no real patterns vertically within the groups. Electronegativities generally increase and then decrease across the d-block periods (left to right) and generally increase further down the groups. The most electronegative d-block metals are therefore generally found towards the bottom right-hand corner of the block (except for Group 12).

Group 7 or VII (17) – the halogens
The halogens consist of: fluorine, F; chlorine, Cl; bromine, Br; iodine, I; and astatine, As. At room temperature, fluorine and chlorine are yellow gases, bromine is a red-brown liquid, iodine is a grey-black solid; and nobody is sure what astatine looks like as it is the rarest naturally occurring element and the heat generated by its radioactivity immediately makes it vaporise. Going down the group:
+ the atomic radius increases
+ ionisation energy decreases
+ electronegativity decreases
+ reactivity decreases
+ melting and boiling points increase.

Group 0 (18) – the noble gases

The noble gases are: helium, He; neon, Ne; argon, Ar; krypton, Kr; xenon, Xe; radon, Rn; and oganesson, Og. All are colourless gases at room temperature, and oganesson is a synthetic element, produced inside a nuclear reactor. All noble gases have completed outer electron shells and as such do not have an electronegativity.

Going down the group:
+ the boiling points increase
+ the atomic radius increases
+ ionisation energy decreases.

Now test yourself   TESTED ◯

15 Complete the following table, summarising the properties going down the groups of the Periodic Table.

	Periodic Table group	Alkali metals	Alkaline earth metals	Halogens	Noble gases
How the property changes down the group	Atomic radius			Increases	
	Ionisation energy		Decreases		
	Electronegativity				Not applicable
	Reactivity				Not applicable

Amount of substance   REVISED ◯

The mole is a convenient way of comparing the amount of substance

Atoms, molecules, ions and electrons are very small, and it is difficult to compare quantities of substance using individual numbers of particles (e.g. 10 molecules of water). Scientists have developed a system of comparing amounts of substance by comparing large quantities of particles that can be weighed or measured. The quantity chosen is called the *mole*. The mole is the name given to a very specific number, 6.023×10^{23}, of atoms, molecules, ions or electrons. Two moles of atoms is 12.046×10^{23} atoms; 0.5 moles of atoms is 3.0115×10^{23} atoms, etc.

In the real world, the amount of substance (in moles) is most closely associated with measurement by mass. The relationship between the mass of a substance and the amount in moles is given by the equation: $\text{moles} = \dfrac{\text{mass}}{M_r}$ where M_r is the relative molecular mass of the substance.

Worked example

Calculate the number of moles of sugar (glucose) in one teaspoon (a mass of 4.2 g).

Formula of glucose = $C_6H_{12}O_6$

Relative molecular mass of glucose, $M_r = (6 \times 12) + (12 \times 1) + (6 \times 16) = 180\,\text{g mol}^{-1}$

$$\text{moles} = \frac{\text{mass}}{M_r} = \frac{4.2\,\text{g}}{180\,\text{g mol}^{-1}} = 0.023\,\text{mol}\,\left(2\,\text{sf}\right)$$

16 Calculate the number of moles of substance in each of the following:
  a 2.3 g of Li
  b 16 g of salt (NaCl) (approx. 4 teaspoons)
  c 47 g of $CaCO_3$
17 Calculate the mass of:
  a 2.5 moles of sulfur
  b 0.4 moles of magnesium oxide, MgO
  c 0.02 moles of sodium sulfate, Na_2SO_4

The mole is used to determine the amount of gas particles

It is quite difficult to weigh gases, but it is relatively easy to measure their pressure, p, volume, V, and temperature, T. These three quantities can be used to determine the amount of gas particles, measured in moles, using the *ideal gas equation*:

$$pV = nRT$$

where **R** is the universal gas constant ($R = 8.3145\, J\, mol^{-1}K^{-1}$).

Exam tip

The ideal gas equation is given on the formulae sheet.

The ideal gas equation is often applied at two specific temperature/pressure combinations, commonly used when calculating the amount of gas particles. When we use these values, the amount of substance only depends on the volume of gas:

+ room temperature and pressure (RTP) – a temperature of 298.15 K (25 °C) and an absolute pressure of exactly 10^5 Pa (100 kPa, 1 bar).
+ standard temperature and pressure (STP) – a temperature of 273.15 K (0 °C) and an absolute pressure of exactly 10^5 Pa (100 kPa, 1 bar).

Maths skills

The ideal gas equation uses *absolute temperatures* measured in kelvin, K. To convert a value in Celsius (°C) to kelvin, use the equation: temperature in K = temperature in Celsius + 273.15.

Worked example

Calculate the number of moles of helium in a balloon with a volume of 0.42 m³ at RTP.

$p = 1.0 \times 10^5$ Pa, $V = 0.42$ m³, $n = ?$, $R = 8.3145\, J\, mol^{-1}K^{-1}$, $T = 298.15$ K (25 °C)

$$n = \frac{pV}{RT} = \frac{1 \times 10^5\, \text{Pa} \times 0.42\, \text{m}^3}{8.3145\, \text{J}\, \text{mol}^{-1}\, \text{K}^{-1} \times 298.15\, \text{K}}$$

$$= 16.94\, \text{moles} = 17\, \text{moles (2 sf)}$$

Practice questions

1 Calculate the volume of 2 moles of oxygen gas at STP.
2 Calculate the number of moles of methane gas in 5.8 m³ at RTP.

43

The amount of substance involved with solutions is linked to the concentration of the solution

Concentration is the amount (or mass) of substance dissolved into a known, standard, volume of a solvent, usually water. The standard volume used by scientists for this measurement is 1 decimetre, 1 dm (also known as 1 litre, 1 l or 1000 cm^3). The amount of substance can be measured in moles (abbreviated to mol, when we are talking about concentrations) or grams, g. Concentration is therefore measured in mol dm^{-3} or g dm^{-3}.

The relationship between the amount of substance, in moles, the volume of a solution, in dm^3, and the concentration, in mol dm^{-3} is given by:
moles = volume (dm^3) × concentration (mol dm^{-3}).

Exam tip

This equation is on the formulae sheet.

Worked example

Calculate the number of moles of sodium hydroxide, NaOH, in 25 cm^3 of 0.50 mol dm^{-3} solution.

$$25\,cm^3 = \frac{25\,cm^3}{1000} = 0.025\,dm^{-3}$$

$$moles = volume\ (dm^3) \times concentration\ (mol\,dm^{-3})$$

$$= 0.025\,dm^{-3} \times 0.50\,mol\,dm^{-3}$$

$$= 0.0125\ moles = 0.013\ moles\ (2\ sf)$$

Now test yourself   TESTED ⬤

18 Calculate the number of moles in 50 cm^3 of 0.12 mol dm^{-3} hydrochloric acid.

19 Calculate the concentration of a 500 cm^3 aqueous solution with 1.8 moles of potassium hydroxide dissolved in it.

The stoichiometry of a formula is the number of moles of each substance involved in a reaction

Chemical formulae are used to show not only the types of reactants and products involved in a reaction, but also the amounts – this is called *stoichiometry*. The numbers before reactants and products in formulae refer to the number of moles of each substance involved in the reaction.

Chemical formulae use the *molecular* or *empirical* formulae of substances. Both types of formula tell you the different types of element involved in the molecule, and the subscripted numbers after the symbols tell you the amount of each element involved. The empirical formula of a molecule is the simplest whole number ratio of atoms of each element in the substance. For example, the molecular formula of glucose is $C_6H_{12}O_6$, but the empirical formula is CH_2O. Empirical formulae are most commonly used to describe the structure of ionic compounds which are often found as giant ionic lattices made up of huge numbers of the ions involved.

Examples of some of the formulae involved with common chemical reactions are as follows.

Acid–base neutralisation

The general formula for acid–base neutralisations is:

$$acid + base \rightarrow salt + water$$

$$H_2SO_4(aq) + 2NaOH(aq) \rightarrow Na_2SO_4(aq) + 2H_2O(l)$$

The stoichiometry of this formula shows that 1 mole of sulfuric acid, H_2SO_4, reacts with 2 moles of sodium hydroxide, NaOH, forming 1 mole of sodium sulfate, Na_2SO_4, and 2 moles of water, H_2O.

Check your understanding and progress at **www.hoddereducation.co.uk/myrevisionnotes**

Thermal decomposition

Thermal decomposition reactions occur when heat is applied to a substance causing it to decompose (or break down) into multiple different chemical substances. Many common thermal decomposition reactions involve metal carbonates. The general formula for the thermal decay of a metal carbonate is:

$$\text{metal carbonate} \xrightarrow{\text{heat}} \text{metal oxide} + \text{carbon dioxide}$$

$$Na_2CO_3(s) \xrightarrow{\text{heat}} Na_2O(s) + CO_2(g)$$

Acid/metal

Some metals react with acids forming a salt and hydrogen gas.

$$\text{metal} + \text{acid} \rightarrow \text{salt} + \text{hydrogen}$$

$$2Li(s) + 2HCl(aq) \rightarrow 2LiCl(aq) + H_2(g)$$

Acid/carbonate

Metal carbonates react with acids to form a metal salt, water and carbon dioxide gas.

$$\text{metal carbonate} + \text{acid} \rightarrow \text{metal salt} + \text{water} + \text{carbon dioxide}$$

$$CaCO_3(s) + 2HCl(aq) \rightarrow CaCl_2(aq) + H_2O(l) + CO_2(g)$$

Precipitation reactions

Two different soluble salts are mixed in solution, forming another soluble salt, and an insoluble salt that forms a white or coloured precipitate.

$$\text{soluble salt A} + \text{soluble salt B} \rightarrow \text{insoluble salt C} + \text{soluble salt D}$$

$$AgNO_3(aq) + KCl(aq) \rightarrow AgCl(s) + KNO_3(aq)$$

Combustion reactions

Most hydrocarbon fuels burn in oxygen (from the air) forming carbon dioxide and water.

$$\text{fuel} + \text{oxygen} \rightarrow \text{carbon dioxide} + \text{water}$$

$$CH_4(g) + 2O_2(g) \rightarrow CO_2(g) + 2H_2O(g)$$

Now test yourself   TESTED

20 Write balanced formulae equations for the following reactions:
   a   Acid/base neutralisation – magnesium oxide reacting with hydrochloric acid.
   b   Acid/carbonate reaction – sodium carbonate reacting with sulfuric acid.
   c   Combustion reaction – propane burning in oxygen.

The formula of a compound determined from experimental data will always be an empirical formula

To calculate the empirical formula of a compound (such as magnesium oxide) formed by the reaction of one substance (e.g. magnesium) with another (e.g. oxygen), we need to calculate the number of moles of each reactant used to create the product, and then convert these numbers to a simple ratio. This is best illustrated by an example.

2.12 g of magnesium reacts with oxygen to form 3.52 g of magnesium oxide. Calculate the empirical formula of magnesium oxide.

First calculate the mass of oxygen used to make the magnesium oxide = 3.52 g – 2.12 g = 1.40 g. Then complete this table.

Element	magnesium	oxygen
Mass used (g)	2.12	1.40
Atomic mass, A_r	24.3	16.0
Number of moles	$\dfrac{2.12\,g}{24.3\,g\,mol^{-1}} = 0.08724\,mol$	$\dfrac{1.40\,g}{16.0\,g\,mol^{-1}} = 0.0875\,mol$
Ratio of moles (÷0.08724)	1	$1.003 \approx 1$
Empirical formula of magnesium oxide	MgO	

Worked example

6.4 g of sulfur reacts with an excess of oxygen, forming 12.8 g of an oxide of sulfur. Calculate the empirical formula of the oxide of sulfur.

Element	sulfur, S	oxygen, O
Mass used (g)	6.4	$12.8\,g - 6.4\,g = 6.4\,g$
Atomic mass, A_r	32.0	16.0
Number of moles	$\dfrac{6.4\,g}{32.0\,g\,mol^{-1}} = 0.20\,mol$	$\dfrac{6.4\,g}{16.0\,g\,mol^{-1}} = 0.40\,mol$
Ratio of moles (÷0.20)	1	2
Empirical formula of magnesium oxide	SO_2	

Now test yourself

TESTED ◯

21 A compound contains 0.32 g of nitrogen and 0.74 g of oxygen. Calculate the empirical formula of the compound.

22 2.60 g of magnesium burns in an excess of chlorine forming 7.60 g of magnesium chloride. Calculate the empirical formula of magnesium chloride.

Calculating reacting masses using correct stoichiometries

Given a correct formula, such as the reaction of sodium burning in an excess of chlorine gas to form solid sodium chloride, it is possible to determine the mass of the product (sodium chloride) given the reacting mass of one of the reactants (e.g. sodium). The correct formula for this reaction is:

$$2Na(s) + Cl_2(g) \rightarrow 2NaCl(s)$$

According to the stoichiometry, 2 moles of sodium, Na, reacts with 1 mole of chlorine gas, Cl_2, forming 2 moles of sodium chloride, NaCl. If 0.52 g of sodium reacts, you can calculate the mass of sodium chloride formed.

Number of moles of Na in 0.52 g = $\dfrac{0.52\,g}{23.0\,g\,mol^{-1}} = 0.0226$ moles

So, number of moles of sodium chloride formed = 0.0226

Molecular mass of sodium chloride = 23.0 + 35.5 = 58.5 g mol⁻¹

0.0226 moles of sodium chloride = 58.5 g mol⁻¹ × 0.0226 moles = 1.3221 g = 1.32 g (3 sf)

> **Worked example**
>
> 6.8 g of phosphorus burns in an excess of oxygen gas forming phosphorus pentoxide, P_2O_5. Write a balanced formula for this reaction and calculate the mass of phosphorus pentoxide formed.
>
> $$4P(s) + 5O_2(g) \rightarrow 2P_2O_5(s)$$
>
> Number of moles of phosphorus in 6.8 g $= \dfrac{6.8\,g}{31.0\,g} = 0.219$ moles $= 0.22$ moles $(2\,sf)$
>
> 4 moles of phosphorus produces 2 moles of phosphorus pentoxide, so
>
> 0.22 moles of phosphorus produces 0.11 moles of phosphorus pentoxide
>
> M_r of phosphorus pentoxide $= (2 \times 31.0) + (5 \times 16.0) = 142\,g\,mol^{-1}$
>
> Mass of 0.11 moles phosphorus pentoxide $= 0.11\,mol \times 142\,g\,mol^{-1} = 15.62\,g$
> $$= 16\,g\,(2\,sf)$$

> **Now test yourself**   TESTED
>
> **23** 3.2 g of calcium burns in an excess of fluorine gas, creating a white powder of calcium fluoride, CaF_2. Write a balanced formula for this reaction and determine the mass of calcium fluoride formed.
>
> **24** Zinc carbonate, $ZnCO_3$, will decompose on heating to form zinc oxide, ZnO, and carbon dioxide, CO_2. If 15 g of zinc carbonate fully decomposes, write a suitable formula for this reaction and calculate the mass of carbon dioxide gas produced.

Acid–base titrations are used to measure concentration

Bases are compounds that react with acids to form *salts*. Soluble bases are called *alkalis*. Reactions between acids and bases are called neutralisation reactions, and a salt and water are formed. Acids have low pH values (less than pH 7), and bases (and alkalis) have high pH values (greater than pH 7).

Strong–weak acids/alkalis

When acids and alkalis dissolve in water, the acidity is determined by the number of free H^+ ions released as the acid molecules dissociate in the water, and alkalinity is determined by the number of OH^- ions that are released as the alkali molecules dissociate. In strong acids/alkalis, all of the acid/alkali molecules dissociate forming H^+/OH^- ions.

Concentrated–dilute acids/alkalis

Concentrated acids/alkalis contain large numbers of H^+/OH^- ions dissolved in the solvent (water). Dilute acids/alkalis contain smaller numbers of H^+/OH^- ions dissolved in the solvent. Even weak acids/alkalis can be concentrated if you have enough acid/alkali molecules.

The pH titration curves of different combinations of acids and alkalis are shown in Figure 1.22.

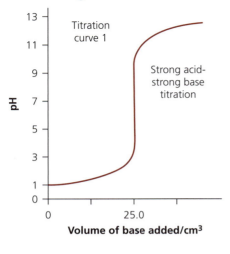

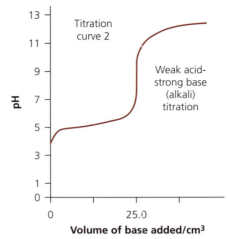

Titration A volumetric way of carrying out a neutralisation reaction. A volumetric pipette measures out a precise volume of acid/alkali and this is titrated against an alkali/acid using a burette and an indicator to find the end point. (Volumetric means 'measured volume'.)

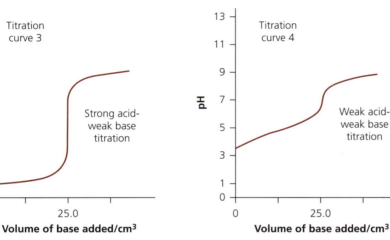

Figure 1.22 pH titration curves

The *equivalence point* of a titration is the vertical section of the pH curve where the volume of alkali (in this case) added causes the pH to rise rapidly. Along the equivalence point, the solution will be neutralised (where pH = 7). The equivalence point refers to a volume of added solution.

The shape of the curve indicates the type of titration (strong/weak acid/alkali). In all the examples in Figure 1.22, an alkali is added to an acid, as the curves start in the acidic portion of the graph. If the curves started with high pH values and curved downwards, this would indicate an acid being added to an alkali.

Worked example

A student titrated $10\,cm^3$ of $0.25\,mol\,dm^{-3}$ hydrochloric acid with dilute sodium hydroxide. She obtained the data shown in Table 1.4 from the titration.

Table 1.4 Results of a titration.

Volume of NaOH added /cm³	0.0	10.0	20.0	24.6	25.0	25.2	25.4	30.0	40.0	50.0
pH	1.0	1.0	1.0	1.1	7.0	12.7	12.9	13.0	13.0	13.0

Making links

You will carry out a titration in Unit 2.

Plot the pH curve for this data and label the equivalence point of the titration.

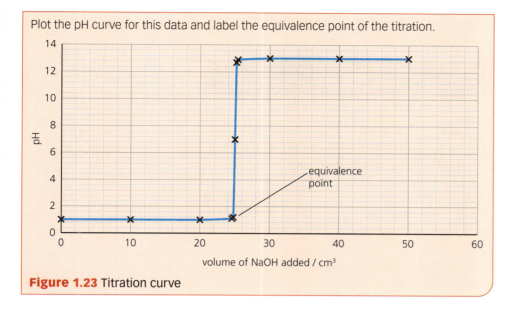

Figure 1.23 Titration curve

Now test yourself

TESTED ⬤

25 Explain the difference between a strong, dilute acid and a weak, concentrated acid.

26 State how you can determine the equivalence point of an acid–base titration.

Choosing an indicator for a titration

An indicator is a substance that changes its colour depending on the pH. Indicators do not take part in the neutralisation reaction itself.

The choice of indicator used in a titration depends on: the types (strengths) of the acids and alkalis involved in the titration; and the resulting pH titration curve. It is obviously important that the indicator changes colour in the pH range of the vertical region (that is, at the equivalence point).

The *end point* of a titration is when the indicator changes colour. This occurs at the equivalence point where there is a rapid change in pH.

The colour change pH ranges of some common indicators are shown in Table 1.5.

Table 1.5 Colour change pH ranges of common indicators

Indicator	pH range of the colour change	Colour in acids	Colour in alkalis
Phenolphthalein	8.3–10.0	Colourless	Pink
Bromothymol blue	6.0–7.6	Yellow	Blue
Methyl red	4.4–6.2	Red	Yellow
Methyl orange	3.1–4.4	Red	Yellow
Thymol blue	1.2–2.8	Red	Yellow

Table 1.6 shows which indicators are best for different types of titration.

Table 1.6 Recommended indicators for different types of titration

Type of titration	pH range of equivalence point	Indicator
Strong acid–strong alkali	3–10	NOT thymol blue
Strong acid–weak base	3–8	NOT thymol blue
Weak acid–strong base	6–10	Bromothymol blue or phenolphthalein
Weak acid–weak base	No vertical region	pH probe must be used

Now test yourself — TESTED

27 Suggest a reason why weak acid–weak base titrations require a pH probe to be used rather than an indicator.

28 Suggest a suitable indicator for a titration involving the weak acid, acetic acid, and the strong base, potassium hydroxide, which has an equivalence point of pH 8.87.

The stoichiometry of a neutralisation formula allows us to calculate the concentrations and/or volumes of unknown solutions

With a titration, we know the volume and concentration of one of the reagents. Knowing this, the stoichiometry, and the equivalence point, we can work out the concentration of the second reagent.

Worked example

$25.0\,cm^3$ of $0.5\,mol\,dm^{-3}$ hydrochloric acid is titrated against an unknown concentration of sodium hydroxide, and the equivalence point is measured to be $37.5\,cm^3$ of sodium hydroxide. Calculate the concentration of the NaOH.

The formula is $HCl(aq) + NaOH(aq) \rightarrow NaCl(aq) + H_2O(l)$

The stoichiometry states that at neutralisation, 1 mole of HCl has reacted with 1 mole of NaOH. The acid reacts with the alkali in a ratio of 1:1. In the earlier worked example we found that the number of moles of HCl present is 0.0125 moles.

Remember: moles = volume (dm^3) × concentration ($mol\,dm^{-3}$)

At the equivalence point, the number of moles of NaOH used = 0.0125 moles, and $37.5\,cm^3$ is $0.0375\,dm^3$, so:

$$\text{Concentration of NaOH} = \frac{0.0125 \text{ moles}}{0.0375\,dm^3} = 0.333\,mol\,dm^{-3} \text{ (3 sf)}$$

Worked example

$25.0\,cm^3$ of $0.2\,mol\,dm^{-3}$ sulfuric acid is titrated against an unknown concentration of sodium hydroxide, and the equivalence point is measured to be $18.6\,cm^3$ of sodium hydroxide. Calculate the concentration of the NaOH.

The formula is

$$H_2SO_4(aq) + 2NaOH(aq) \rightarrow Na_2SO_4(aq) + 2H_2O(l)$$

The stoichiometry states that at neutralisation, 1 mole of H_2SO_4 has reacted with 2 moles of NaOH. The acid reacts with the alkali in a ratio of 1:2.

$$\text{Number of moles of } H_2SO_4 \text{ present} = \frac{25.0\,cm^3}{1000} \times 0.2\,mol\,dm^{-3} = 0.005 \text{ moles}$$

At the equivalence point, the number of moles of NaOH used = 2 × 0.005 = 0.010 moles, so:

$$\text{Concentration of NaOH} = 0.010 \text{ moles} \times \frac{1000\,cm^3}{18.6\,cm^3} = 0.538\,mol\,dm^{-3} \left(3\,sf\right)$$

Check your understanding and progress at **www.hoddereducation.co.uk/myrevisionnotes**

Now test yourself   TESTED ⬤

29 25.0 cm³ of 0.12 mol dm⁻³ hydrochloric acid is titrated against an unknown concentration of lithium hydroxide, LiOH, and the equivalence point is measured to be 28.9 cm³ of lithium hydroxide. Calculate the concentration of the lithium hydroxide.

30 25.0 cm³ of 3.5 mol dm⁻³ sulfuric acid is titrated against an unknown concentration of potassium hydroxide, KOH, and the equivalence point is measured to be 15.6 cm³ of potassium hydroxide. Calculate the concentration of the potassium hydroxide.

> **Exam tip**
>
> To convert a volume in cm³ to dm³, divide by 1000.

Bonding and structure

REVISED ⬤

The properties of a substance are dictated by the types of atoms involved and the structure of the substance. There are three main types of bonding: ionic, covalent and metallic. By observing the properties of the substance, it is possible to predict the type of bonding involved.

> **Exam tip**
>
> You may be required to identify the structure of an unknown substance from its properties.

Ions and ionic compounds

Ions are formed due to the loss or gain of electrons. Cations are formed when there is a loss of electrons. Cations are positively charged and are attracted to a cathode when molten or in aqueous solution. Anions are formed when there is a gain of electrons. Anions are negatively charged and are attracted to the anode when molten or in aqueous solution.

Cations are generally formed when metal atoms lose electrons: alkali metals lose one electron (e.g. Li^+; Na^+; K^+); alkaline earth metals lose two electrons (e.g. Mg^{2+}; Ca^{2+}); transition metals generally lose two or three electrons (e.g. Cu^{2+}; Fe^{2+}; Fe^{3+}; Zn^{2+}; Ni^{2+}); and aluminium loses three electrons, Al^{3+}.

Anions are formed when non-metals or non-metal complexes gain electrons: halogens gain one electron (e.g. Cl^-; Br^-; I^-); oxygen gains two electrons, O^{2-}.

The common complex anions are shown in Table 1.7.

Table 1.7 Common complex anions

Anion	sulfate	carbonate	nitrate	hydroxide
Electrons gained	2	2	1	1
Formula	SO_4^{2-}	CO_3^{2-}	NO_3^-	OH^-

The charges on ions can be used to deduce the formulae of ionic compounds because the positive and negative charges must balance out so that the ionic compound is overall neutral.

> **Worked example**
>
> Deduce the formula of aluminium carbonate.
>
> Aluminium ions have a 3+ charge: Al^{3+}
>
> Carbonate ions have a 2– charge: CO_3^{2-}
>
> $2 \times Al^{3+}$ ions are needed to balance the charge of $3 \times CO_3^{2-}$ ions
>
> The formula of aluminium carbonate is $Al_2(CO_3)_3$.

Now test yourself   TESTED ⬤

31 Deduce the formulae of the following ionic compounds:

   a lithium sulfate
   b calcium nitrate
   c iron(III) carbonate
   d magnesium hydroxide.

There are several types of bonding

Ionic bonding

Ionic bonds are formed between the ions in ionic compounds. The electrostatic forces of attraction between the ions are strong. The ionic bonding between the ions means that they form giant ionic crystal lattices. These are three-dimensional, regular repeating arrays of positive and negative ions. A generalised ionic lattice is shown in Figure 1.24. This represents a similar structure to the ionic lattice in sodium chloride.

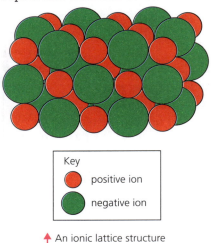

Key
⬤ positive ion
🟢 negative ion

↑ An ionic lattice structure

Figure 1.24 Each Na+ ion is surrounded by six Cl- ions and each Cl- ion is surrounded by six Na+ ions. The regular structure continues like this to form the crystal.

In this case, each sodium ion is surrounded by six chlorine ions as shown in Figure 1.25.

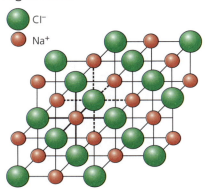

🟢 Cl-
🔴 Na+

Figure 1.25 Sodium chloride lattice of ions

Magnesium oxide has a similar structure to sodium chloride.

Covalent bonding

A covalent bond is formed when two non-metals share one or more pairs of electrons. Covalent bonds are usually illustrated using *dot and cross diagrams*, which show how the outer electrons are arranged in a bond. The electrons of one atom are shown as crosses; the electrons of the other atoms are shown as dots. A shared pair of electrons is represented by a dot and a cross (· ×) to show that the electrons in the bond come from different atoms.

(Simple) Covalent molecules are formed when a non-metal atom makes a covalent bond with another non-metal atom, or with several other non-metal atoms.

Iodine forms diatomic molecules. The dot and cross diagram for an iodine molecule, together with the bonding diagram (where the bond is represented by a single line), are shown in Figure 1.26.

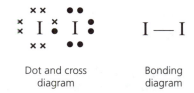

Dot and cross diagram   Bonding diagram

Figure 1.26 A dot and cross diagram and a bonding diagram for an iodine molecule

Check your understanding and progress at **www.hoddereducation.co.uk/myrevisionnotes**

Methane (CH_4) is formed by four covalent bonds between the four outer electrons of carbon and the outer electron of four hydrogen ions. The dot and cross diagram and the bonding diagram for methane are shown in Figure 1.27.

Dot and cross diagram   Bonding diagram

Figure 1.27 Dot and cross diagram and bonding diagram for methane, CH_4

Covalent bonds can also be double bonds (or even triple bonds). Double bonds are formed when two atoms share four electrons. The bonds between the carbon and oxygen atoms in carbon dioxide molecules form double bonds, as shown in Figure 1.28.

Dot and cross diagram   Bonding diagram

Figure 1.28 Double bonding in a carbon dioxide molecule

The interatomic covalent bonds are very strong, but the intermolecular (between molecules) forces of attraction are very weak.

Metallic bonding

When metal atoms coalesce to form a solid, the outer electrons become de-localised from individual atoms and the remaining positive ion cores form a three-dimensional lattice. The de-localised electrons are free to move throughout the structure of the solid metal and behave like a conducting gas. A two-dimensional representation of the metallic bonding in magnesium is shown in Figure 1.29.

Alloys are mixtures of metals. There are two forms of alloy structure: substitutional alloys, where atoms of one metal are replaced with atoms of a second metal, and interstitial alloys, where smaller metal atoms fit into the spaces between the larger atoms. These two structures are shown in Figure 1.30.

Magnesium

Figure 1.29 Metallic bonding in magnesium

Substitution alloy

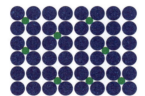

Interstitial alloy

Figure 1.30 Alloy structures

Alloys have physical properties that can be changed by the addition of different amounts and types of each metal. Materials scientists have become very skilled at creating alloys for different uses. Brass, for example, is an alloy of copper and zinc. By varying the proportions of the different metals, scientists can manufacture brass to have a wide range of different properties, for a wide range of uses. Brass does not corrode, so it is ideal for making metal fittings that stay outside, or in marine environments. A different composition of brass is used to make musical instruments and fixing screws.

> **Exam tip**
>
> Questions may ask you to draw the structure of a substance. You should learn the general structural diagrams of ionic lattices, simple covalent molecules, covalent macromolecules and metals.

53

The type of bonding dictates the physical properties of a substance

Table 1.8 Physical properties of substances with different types of bonding

Property	Ionic	Covalent	Metallic
Electrical conductivity	Poor when solid because although the structure contains charged ions, these are fixed in position. Good when molten as the charged ions are free to move.	Generally, electrically neutral. This means there are no charged particles free to move and conduct electricity.	High due to delocalised electrons.
Melting and boiling points	High because the forces of electrostatic attraction between the ions are very strong and so a great deal of energy is required to break them.	Generally low because the intermolecular forces of attraction are weak and so little energy is needed to overcome them.	Generally high due to strong attractive forces between the positive ion cores and the delocalised electrons.
	Low because the forces of electrostatic attraction between the ions are very strong.	Generally high because the intermolecular forces are weak, so little energy is needed to break apart the molecules when they are close-packed in a liquid. Covalent molecules are generally gases at room temperature and pressure.	Non-volatile
Solubility in water	Generally soluble because water is a **polar** molecule and tends to surround ions that have broken off the lattice due to the water molecules knocking ions off the structure. In some ionic compounds, such as aluminium oxide, the ionic bonds are so strong that water cannot break up the lattice and the compound is insoluble in water.	Generally insoluble in water, because covalent molecules are non-polar.	Insoluble
Solubility in non-polar solvents	Generally insoluble in non-polar solvents, such as hexane, as the intermolecular forces of the solvent are generally weaker than the bonds between the ions.	Generally soluble in non-polar solvents such as hexane.	Insoluble

Polar Some overall neutral molecules can be described as polar molecules. The shared electrons in the bonds tend to be closer to one of the atoms than to the others. As a result, one end of the molecule becomes slightly negatively charged and the other end becomes slightly positively charged.

Volatility How readily a substance vaporises. Volatility depends upon temperature and pressure, but a substance with a high volatility is more likely to exist as a vapour (gas), whereas a substance with a low volatility is more likely to be a liquid or solid.

Check your understanding and progress at **www.hoddereducation.co.uk/myrevisionnotes**

Giant covalent structures and carbon

Giant covalent structures are formed when non-metal atoms combine with covalent bonds but instead of forming simple molecules, the structure forms a giant three-dimensional lattice. These structures usually involve carbon or silicon. Carbon forms four different types of giant covalent structure: diamond, graphite, graphene and fullerene. Different physical forms of the same element are called *allotropes*. The structure of each allotrope dictates its properties.

Table 1.9 Allotropes of carbon

	Diamond	Graphite	Graphene	Fullerene
Bond	Four strong single covalent bonds with four other carbon atoms.	Three strong covalent bonds with three other carbon atoms arranged in a hexagonal layer and a weak, delocalised covalent bond between adjacent layers of carbon atoms.	Individual layers of graphite are called *graphene*. Graphene is incredibly thin and has delocalised electrons which are free to move across its surface. Graphene can be arranged as a sheet or as tubes (frequently called carbon nanotubes) and is very strong (200 times stronger than steel).	Fullerenes form when sheets of graphene roll up into cages. One example, Buckminsterfullerene, contains 60 carbon atoms.
Diagram		Mobile electrons between the layers	Carbon nanotube Graphene	
Melting and boiling points	High as covalent bonds are strong.	High as bonds between layers are strong.	Very high as covalent bonds are strong.	Low as there are only weak intermolecular forces.
Electrical conductivity	No free charged particles so do not conduct electricity.	The forces of attraction holding the atoms within the layers are very strong. However, the forces of attraction between the layers are very weak, and the electrons forming this bond are delocalised and free to move throughout the gap between the layers. Conducts electricity in the direction of the layers.	Excellent conductors of electricity.	Fullerenes conduct electricity within the molecules, but not from molecule to molecule, so they are generally considered to be insulators.
Volatility	Not volatile	Not volatile	Very low volatility	Volatility decreases with increasing cage size
Solubility in water	Insoluble	Insoluble	Insoluble	Only slightly soluble in water
Solubility in non-polar solvents	Insoluble	Insoluble	Insoluble	Generally insoluble

Enthalpy changes

REVISED

Enthalpy, exothermic and endothermic reactions

Reactions that release heat to the surroundings (so that the reaction vessel feels warmer) are called *exothermic* reactions. Reactions that absorb heat from the surroundings (so that the reaction vessel feels colder) are called *endothermic* reactions.

All substances have an internal energy. This energy is associated with the bonding between the atoms in the substance. Each type of bond has a characteristic energy associated with it and the total sum of all the energies associated with a substance is called its *enthalpy*, and is given the symbol H. The enthalpy of a substance tells you how much energy is needed to break all the bonds (per mole of substance) or it tells you how much energy is needed to make all the bonds in the substance (per mole of substance).

If the enthalpy of the reactants in a reaction is given as H_1, and the enthalpy of the products is given as H_2, then the change of enthalpy, ΔH, is given by $\Delta H = H_2 - H_1$.

When ΔH is negative, the reaction is exothermic, and when ΔH is positive the reaction is endothermic. Enthalpy changes are measured in kilojoules per mole of substance, so the unit is $kJ\,mol^{-1}$.

There are several types of enthalpy change. All enthalpy changes are given under standard conditions (STP), i.e. 273.15K (0°C) and 1×10^5 Pa. The enthalpy of reaction, $\Delta_{rx}H$, is the enthalpy change when substances react in quantities shown by the formula for the reaction.

For example:

$$2H_2(g) + O_2(g) \rightarrow 2H_2O(l) \qquad \Delta_{rx}H = -572\,kJ\,mol^{-1}$$

When 2 moles of hydrogen gas react with 1 mole of oxygen forming 2 moles of water, 572kJ of thermal energy is released to the surroundings. At constant pressure, the change in enthalpy is equal to the thermal energy released (or absorbed) by the reaction. As a result, the change in enthalpy is often referred to as the heat of reaction.

The enthalpy of formation, $\Delta_f H$, is the enthalpy change when 1 mole of substance is formed from its constituent elements. The formula above, showing the reaction of hydrogen and oxygen, ends up forming 2 moles of water. To calculate the enthalpy of formation of water, the reaction needs to be shown as:

$$H_2(g) + \frac{1}{2}O_2(g) \rightarrow H_2O(l) \qquad \Delta_{rx}H = -286\,\text{kJ mol}^{-1}$$

In this formula, 1 mole of hydrogen gas reacts with half a mole of oxygen gas, forming 1 mole of water.

The enthalpy of neutralisation is a special sub-category of the enthalpy of formation as it describes change in enthalpy when an acid and base undergo a neutralisation reaction to form 1 mole of water.

The enthalpy of combustion, $\Delta_c H$, is the enthalpy change when 1 mole of a substance burns completely in an excess of oxygen. In the example above, where hydrogen burns in oxygen, the enthalpy of formation of water is the same value as the enthalpy of combustion of hydrogen.

Enthalpies are frequently illustrated by drawing energy profiles. The generalised energy profiles for an exothermic and an endothermic reaction are shown in Figures 1.31 and 1.32.

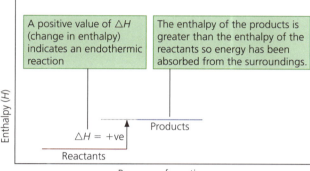

The value of the enthalpy change (ΔH) in this endothermic reaction is positive. This is because there has been an increase in enthalpy from reactants to products. This is a standard feature of **endothermic** reactions and it must be remembered that **ΔH is positive.**

Figure 1.31 An enthalpy level diagram for an endothermic reaction

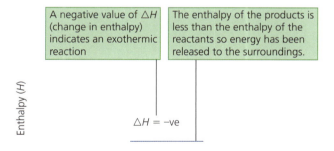

The value of the enthalpy change (ΔH) in this exothermic reaction is negative. This is because there has been a decrease in enthalpy from reactants to products. This is a standard feature of **exothermic** reactions and it must be remembered that **ΔH is negative.**

Figure 1.32 An enthalpy level diagram for an exothermic reaction

---

Now test yourself   TESTED ⬤

39 Explain the difference between the enthalpy of reaction of hydrogen burning in oxygen to form water; the enthalpy of formation of water; and the enthalpy of combustion of hydrogen

Energy profiles and activation energy

Although the energy profiles shown in Figure 1.31 are useful, they do not show the pathway, in terms of energy, from reactants to products. Reactions require an activation energy to break the bonds in the reactants, before the new bonds form in the products. The activation energy is shown as a hump in the energy profile, and the height of the hump above the enthalpy of the reactants gives the size of the activation energy, as shown in Figures 1.33 and 1.34.

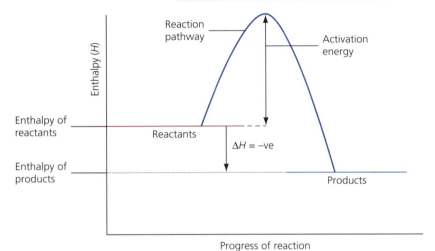

This shows that the activation energy is the minimum amount of energy which the reactants must have in order to react. Some reactions have low activation energy and can obtain enough energy at room temperature to raise the reactants to the required enthalpy value to allow the reaction to proceed.

Figure 1.33 The reaction pathway for an exothermic reaction

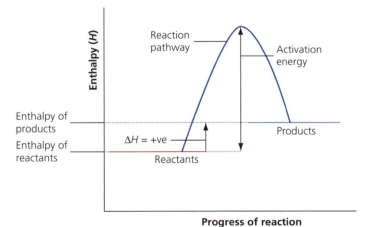

This shows that the activation energy is the amount of energy which the reactants must have in order to react. Many endothermic reactions have a high activation energy and cannot obtain enough energy at room temperature to raise the reactants to the required enthalpy value to allow the reaction to proceed.

Figure 1.34 The reaction pathway for an endothermic reaction

40 The activation energy for the combustion of methane in oxygen is $+2650\,kJ\,mol^{-1}$. The enthalpy of reaction is $-890\,kJ\,mol^{-1}$. Draw the energy profile reaction pathway for this reaction.

Check your understanding and progress at **www.hoddereducation.co.uk/myrevisionnotes**

Calculating enthalpy changes using Hess's Law

Hess's Law is an expression of the law of conservation of energy. It states that the enthalpy change of a chemical reaction only depends on the initial enthalpy of the reactants and the final enthalpy of the products: it is independent of the route of the reaction (the activation energy).

Hess's Law can be used to calculate the (theoretical) enthalpy change of a chemical reaction. There are two ways to do this: by using a Hess's Law cycle diagram or by using a sum of the enthalpy changes.

When using a Hess's Law cycle diagram, the key is to remember that the reaction pathway does not matter. It is possible to use enthalpies of combustion (which are easy to measure experimentally) to calculate enthalpies of formation; and enthalpies of formation to calculate enthalpies of reaction. Figure 1.35 shows how to calculate the energy of formation of a compound from enthalpies of combustion.

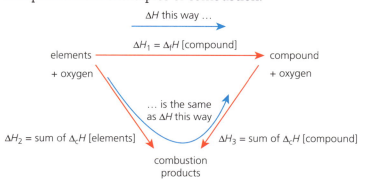

Figure 1.35 An energy cycle diagram for calculating standard enthalpies of formation from standard enthalpies of combustion: $\Delta H_1 = \Delta H_2 - \Delta H_3$

Worked example

Calculate the enthalpy of formation of methane.

$$C(s) + 2H_2(g) \rightarrow CH_4(g)$$

The enthalpies of combustion involved with this reaction are:

$$C(s) + O_2(g) \rightarrow CO_2(g) \qquad \Delta_c H = -393 \, kJ \, mol^{-1}$$

$$H_2(g) + \frac{1}{2}O_2(g) \rightarrow H_2O(l) \qquad \Delta_c H = -286 \, kJ \, mol^{-1}$$

$$CH_4(g) + 2O_2(g) \rightarrow CO_2(g) + 2H_2O(l) \quad \Delta_c H = -890 \, kJ \, mol^{-1}$$

The Hess's Law cycle diagram for this reaction is shown in Figure 1.36.

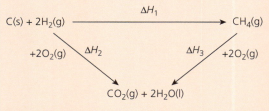

Figure 1.36 Hess's Law cycle diagram for the formation of methane

Hess's Law states: $\Delta H_1 = \Delta H_2 - \Delta H_3$

ΔH_1 is the enthalpy of formation, $\Delta_f H$, required in the question.

$$\Delta H_2 = -393 \, kJ \, mol^{-1} + (2 \times -286) = -965 \, kJ \, mol^{-1}$$

$$\Delta H_3 = -890 \, kJ \, mol^{-1}$$

$$\Delta_f H = \Delta H_1 = (-965 \, kJ \, mol^{-1}) - (-890 \, kJ \, mol^{-1}) = -75 \, kJ \, mol^{-1}$$

Figure 1.37 shows the general Hess's law cycle diagram for calculating enthalpies of reaction from enthalpies of formation.

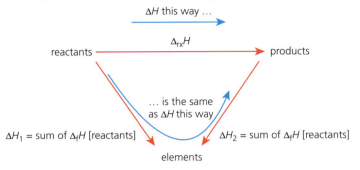

Figure 1.37 An energy cycle for calculating standard enthalpies of reaction from standard enthalpies of formation

---

Worked example

Calculate the enthalpy of reaction of the reaction of iron(III) oxide with carbon monoxide to form iron.

$$Fe_2O_3(s) + 3CO(g) \rightarrow 2Fe(s) + 3CO_2(g)$$

The enthalpies of formation, $\Delta_f H$, are:

$Fe_2O_3 = -824\,kJ\,mol^{-1}$

$CO = -110\,kJ\,mol^{-1}$

$Fe = 0$

$CO_2 = -393\,kJ\,mol^{-1}$

The Hess's Law cycle diagram for this reaction is shown in Figure 1.38.

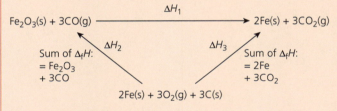

Figure 1.38 Hess's Law cycle diagram for the formation of iron

ΔH_1 is the enthalpy of reaction, $\Delta_{rx}H$, $= -\Delta H_2 + \Delta H_3 = \Delta H_3 - \Delta H_2$

$\Delta_{rx}H = (3 \times -393\,kJ\,mol^{-1}) - (-824\,kJ\,mol^{-1} + (3 \times -110\,kJ\,mol^{-1}))$
$= (-1179 + 1154)\,kJ\,mol^{-1} = -25\,kJ\,mol^{-1}$

41 Butane gas can be formed by reacting carbon with hydrogen. The formula for this reaction is:

$$4C(s) + 5H_2(g) \rightarrow C_4H_{10}(l)$$

Calculate the enthalpy of formation of butane.

The enthalpies of combustion for this formula are:

Carbon $-394\,kJ\,mol^{-1}$

Hydrogen $-286\,kJ\,mol^{-1}$

Butane $-2878\,kJ\,mol^{-1}$

42 Magnesium oxide reacts with hydrogen to form magnesium metal and water. The formula for this reaction is:

$$MgO(s) + H_2(g) \rightarrow Mg(s) + H_2O(l)$$

Calculate the enthalpy of reaction for this reaction.

The enthalpies of combustion for this formula are:

Magnesium $-602\,kJ\,mol^{-1}$

Hydrogen gas $-242\,kJ\,mol^{-1}$

Calculating enthalpy changes using mean bond enthalpies

When calculating enthalpy changes using the sum of mean bond enthalpies, the first step is to determine the enthalpy sum of breaking all the bonds in the reactants (these will be negative values). The second step is to calculate the enthalpy sum of making all the bonds in the products (these will be positive values). The last step is to determine the change in the enthalpy sums.

Mean bond enthalpies are average values because bond enthalpies vary slightly from one compound to another, and if a molecule involves more than one bond of the same type, breaking (or making) the first bond requires a different amount of energy than the second bond.

Worked example

Calculate the enthalpy of formation of water formed by the combustion of hydrogen in oxygen. The reaction is:

$$2H_2(g) + O_2(g) \rightarrow 2H_2O(l)$$

Remember that the enthalpy of formation involves creating 1 mole of the product, so the final value from the calculation will need to be halved.

The first step is to break all the bonds in the hydrogen and oxygen molecules, and then make all the bonds in the water molecules. The energy profile for this reaction, showing the atoms and molecules involved, is shown in Figure 1.39.

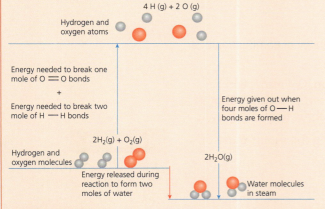

Figure 1.39 An energy level diagram for the reaction between hydrogen and oxygen

Bonds broken	Enthalpy /kJ mol^{-1}	Bonds made	Enthalpy /kJ mol^{-1}
2 × H–H	+ (2 × 436)	4 × O–H	– (4 × 464)
1 × O=O	+ 498		
Total	+1370		–1856

$\Delta H = +1370\,\text{kJ mol}^{-1} - 1856\,\text{kJ mol}^{-1} = -486\,\text{kJ mol}^{-1}$

So, $\Delta_f H = \dfrac{-486\ \text{kJ mol}^{-1}}{2} = -243\ \text{kJ mol}^{-1}$

Now test yourself

TESTED ◯

43 Calculate the enthalpy of reaction of the combustion of ethene gas, C_2H_4. The formula for this reaction is:

$$C_2H_4(g) + 3O_2(g) \rightarrow 2CO_2(g) + 2H_2O(l)$$

Bond enthalpies:

C=C 619 kJ mol^{-1}

C–H 414 kJ mol^{-1}

O=O 499 kJ mol^{-1}

C=O 724 kJ mol^{-1}

O–H 464 kJ mol^{-1}

Measuring enthalpy changes

Measuring enthalpies in the laboratory generally involves measuring a temperature change of water (or an aqueous solution).

Figure 1.40 shows the experimental arrangement needed to measure the enthalpy of combustion of a liquid fuel such as ethanol.

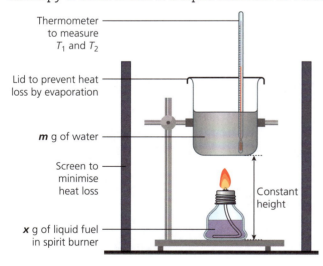

Thermometer to measure T_1 and T_2

Lid to prevent heat loss by evaporation

m g of water

Screen to minimise heat loss

Constant height

x g of liquid fuel in spirit burner

Figure 1.40 Experimental arrangement to measure the enthalpy of combustion of a liquid fuel

The method relies on the specific heat capacity equation, $Q = mc\Delta T$, where Q is the amount of thermal energy needed to raise the temperature, ΔT, of a known mass, m, of a material (usually water) with a specific heat capacity, c.

The number of moles of fuel (ethanol) is calculated using the molar mass of the fuel, M_r:

$$\text{moles} = \frac{\text{mass of fuel, } g}{\text{molar mass, } M_r}$$

Making links

The concept of specific heat capacity is covered in Unit 2.

Check your understanding and progress at **www.hoddereducation.co.uk/myrevisionnotes**

In Figure 1.40, m grams of water is heated by x grams of ethanol in a spirit burner. The temperature change of the water, ΔT, is calculated using the start temperature, T_1, and the end temperature, T_2 ($\Delta T = T_2 - T_1$). The mass of the water, m, needs to be converted to kilograms, and the specific heat capacity of water is $4.186\,kJ\,kg^{-1}\,°C^{-1}$. The energy absorbed by the water, Q (in kJ), is then divided by the number of moles to convert it to $kJ\,mol^{-1}$.

A similar method can be used to measure the *enthalpy of neutralisation* of an acid–base reaction.

Consider the neutralisation of hydrochloric acid with sodium hydroxide:

$$HCl(aq) + NaOH(aq) \rightarrow NaCl(aq) + H_2O(l)$$

The stoichiometry states that 1 mole of hydrochloric acid reacts with 1 mole of sodium hydroxide, forming 1 mole of water and 1 mole of sodium chloride. If $25\,cm^3$ of $1.0\,mol\,dm^{-3}$ hydrochloric acid is added to $25\,cm^3$ of $1.0\,mol\,dm^{-3}$ sodium hydroxide there will be complete neutralisation. This reaction is exothermic and the temperature of the reaction vessel, including the water, will increase. The enthalpy of neutralisation can be calculated by measuring the temperature change of the (known) volume of the solution (mostly water).

The temperature of the reaction is monitored over time and a graph plotted of the temperature against the time. A correction technique is used to determine the maximum temperature because the solution does not heat evenly and some energy is lost due to thermal transfer to the surroundings. The correction technique is shown in Figure 1.41.

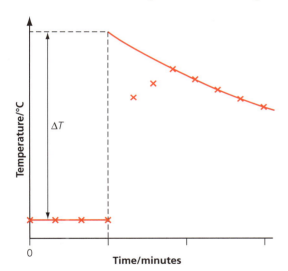

Figure 1.41 Estimating the maximum temperature of a neutralisation reaction

The maximum temperature is estimated using the cooling portion of the graph and extrapolating the cooling curve back to the time of mixing the acid and the alkali. The mass of the water is either measured directly (using a balance) or calculated by using the density of water. The enthalpy of neutralisation can then be calculated by using the specific heat capacity equation, as (in this case) both solutions are $1.0\,mol\,dm^{-3}$.

Now test yourself   TESTED

44 Draw a labelled diagram of the apparatus that you would need to calculate the enthalpy of neutralisation of $0.1\,mol\,dm^{-3}$ hydrochloric acid with $0.1\,mol\,dm^{-3}$ sodium hydroxide solution.

Summary

+ Atoms consist of protons, neutrons and electrons.
+ The atomic (proton) number has the symbol Z and the mass number has the symbol A.
+ Isotopes have the same number of protons but different numbers of neutrons and different elements have different isotopic abundances.
+ Elements have different electron sub shell configurations; electron transitions give rise to emission spectra and compound colour.
+ Relative atomic, molecular and formula masses are given in terms of $^{12}_{6}C$.
+ The Periodic Table lists elements in proton number order; arranged in periods/rows and groups/columns.
+ The Periodic Table arranges elements in s, p and d-blocks with different patterns of: atomic radii, ionisation energy and electronegativity.
+ A mole of any substance contains the same number of particles; and:

$$\text{moles} = \frac{\text{mass}}{\text{atomic or molar or formula mass}}$$

+ For a gas, $pV = nRT$.
+ Molecules have molecular formulae; ionic compounds have empirical formulae.
+ Balanced equations allow the reacting masses to be calculated using the correct stoichiometry.
+ pH curves can be used to determine the equivalence point of an acid–base titration.
+ The choice of titration indicator depends on the strength of the acid and base used and the pH titration curve.
+ In volumetric analysis: moles = volume (dm^3) × concentration ($mol\,dm^{-3}$).
+ Ionic bonds involve exchange of electrons forming strong ionic lattices.
+ Covalent bonds: share pairs of electrons; form neutral molecules; are non-conductors; and have weak intermolecular forces.
+ Giant covalent structures contain lattices of covalent bonds.
+ Metallic bonding involves a strong metallic lattice with free electrons.
+ Key material properties are dictated by the type of bonding.
+ Energy profiles can be used to show exothermic and endothermic reactions and the activation energy of a reaction.
+ Mean bond enthalpies and Hess's Law cycles can be used to calculate enthalpy changes.
+ The molar enthalpy of combustion of a liquid fuel, and the molar enthalpy of neutralisation, can be determined practically, using: $Q = mc\Delta T$.

Exam practice

1 Iron(III) oxide, Fe_2O_3, is a red-brown compound found as the mineral hematite. Hematite is the main ore of iron, used for the production of steel.

   1.1 Calculate the relative formula mass of iron(III) oxide. [3]

   1.2 Explain why iron(III) ions are red-brown in colour. [2]

   1.3 Iron has four naturally occurring isotopes: iron-54, iron-56, iron-57 and iron-58. Explain what is meant by an isotope. [1]

   1.4 The isotopic abundances of the four naturally occurring iron isotopes are shown in the table.

Isotope	Abundance (%)
iron-54	5.85
iron-56	91.75
iron-57	2.12
iron-58	0.28

   Calculate the relative atomic mass of iron. Give your answer to 3 significant figures [3]

2 Lithium, sodium and potassium are three elements commonly found in school and college chemical stores. They are frequently used to illustrate how the properties of elements change going down a Periodic Table group.

   2.1 State the property of elements that is used to arrange the Periodic Table in order. [1]

   2.2 State the name of the group and the block that contains lithium, sodium and potassium. [2]

   2.3 Explain why lithium, sodium and potassium all react vigorously with water, producing hydrogen gas and a hydroxide solution. [2]

   2.4 Explain why lithium has a higher ionisation energy than sodium. [2]

Check your understanding and progress at **www.hoddereducation.co.uk/myrevisionnotes**

2.5 The table shows the element symbol; the number of protons, Z, inside the nucleus and the atomic radius, r_a, of each element in the group.

Symbol	Li	Na	K	Rb	Cs	Fr
Z	3	11	19	37	55	87
r_a (×10^{-10} m)	1.5	1.9	2.3	2.5	2.7	3.5

    2.5.1 Plot these data on graph paper. Draw a line of best fit. [2]

    2.5.2 State and explain the trend shown in the graph. [3]

3 Magnesite (magnesium carbonate, $MgCO_3$), is a white powder frequently used in dry fire extinguishers. Magnesite reacts with hydrochloric acid forming a salt, water and carbon dioxide gas.

3.1 Identify the salt produced by this reaction. [1]

3.2 20.0 g of magnesite is used in the reaction, together with an excess of acid. Calculate the mass of the magnesium in the magnesite. [3]

3.3 22.6 g of the salt are produced. Calculate the empirical formula of the salt. [3]

3.4 Write a balanced symbol equation for the reaction of magnesite and hydrochloric acid. Include the state symbols. [3]

3.5 Identify the type of reaction. [1]

    A Neutralisation   **D** Acid/carbonate

    B Combustion   **E** Acid/metal

    C Thermal decomposition   **F** Precipitation

3.6 The reaction produces 0.0057 m³ of carbon dioxide gas at RTP. Calculate the number of moles of carbon dioxide gas produced. The gas constant, $R = 8.31\,J\,K^{-1}\,mol^{-1}$. [3]

4 A technician is asked to determine the concentration of an unknown acid water sample. The water has been contaminated by hydrochloric acid. They titrate 25 cm³ of 0.1 mol dm⁻³ sodium hydroxide against the contaminated acidic water. They measure the pH as the acidic water is added using a burette. Their results are shown in the table.

Volume of acid added/cm³	0.0	4.0	8.0	12.0	15.8	16.4	18.0	20.0	22.0	24.0
pH	11.2	10.9	10.7	10.3	8.9	2.6	1.9	1.7	1.5	1.3

4.1 Plot a graph of these results and draw a suitable line of best fit. [2]

4.2 Use your graph to determine the equivalence point of this reaction. [1]

4.3 Write a balanced symbol equation for this reaction. Include the state symbols. [3]

4.4 Identify this type of neutralisation reaction: [1]

    A Strong acid–strong alkali   **C** Weak acid–strong alkali

    B Strong acid–weak alkali   **D** Weak acid–weak alkali

4.5 Suggest a suitable indicator for this reaction choosing from the indicators listed in the table on page 49. [1]

4.6 Calculate the concentration of the acidic water. [3]

5 Cubic zirconia (ZrO_2) is an oxide of the element zirconium. It is a hard, shiny, transparent ionic compound that is frequently used to make cheap jewellery because of its physical similarities to diamond.

5.1 State what is meant by an ionic compound. [2]

5.2 Calculate the charge on a zirconium ion. [2]

5.3 Explain why cubic zirconia will only conduct electricity at high temperatures. [2]

5.4 Diamond is an allotrope of carbon. Explain what is meant by an allotrope. [1]

5.5 Diamond is a giant covalent macromolecule. Draw a labelled diagram of the structure of diamond. Your diagram should be a 3-dimensional representation. [2]

5.6 Explain why diamond has a high melting and boiling point. [2]

5.7 Zirconium is a transition metal with a structure similar to magnesium. Describe the structure of zirconium. [2]

5.8 Oxygen is a diatomic covalent molecule, involving a double covalent bond. Draw a dot and cross diagram of an oxygen molecule. [2]

Unit 1 Key concepts in the application of science

6. A science technician wants to measure the enthalpy of combustion of ethanol. They propose to do this by heating water using an ethanol spirit burner

   6.1 State the two key pieces of **measuring** apparatus that the technician will need to use to perform this experiment. [2]

   6.2 Explain why any value for the enthalpy of combustion determined from this experiment will be lower than the true value. [2]

   6.3 Write a balanced formula for the combustion of ethanol, C_2H_5OH. [3]

   6.4 Calculate the enthalpy of combustion of ethanol. Give your answer to 3 significant figures. [4]

   The bond enthalpies for this reaction are given below:

C–C	$346\,kJ\,mol^{-1}$
C–H	$414\,kJ\,mol^{-1}$
C–O	$358\,kJ\,mol^{-1}$
O=O	$499\,kJ\,mol^{-1}$
C=O	$724\,kJ\,mol^{-1}$
O–H	$464\,kJ\,mol^{-1}$

Key concepts in the application of physics

Introduction

REVISED

In this section you will find out how to compare the efficiency of different devices, and apply the concept of power – the rate of energy transferred by a device. You will learn how U-values are used by energy consultants and architects to measure the effectiveness of different materials as building insulators, reducing heat loss. You will also investigate the environmental concerns associated with fossil fuels and nuclear methods of producing electricity, which have given rise to the use of alternative methods for the generation of energy.

Most devices contain electric circuits. You will investigate how the properties of an electrical circuit are altered by adding different components.

You will also discover how Newton's laws of motion are used to predict the motion and interaction of objects, particularly in the context of cars and sport.

Useful energy and efficiency

REVISED

Maximising efficiency reduces wasted energy and improves economy

The efficiency of devices and processes is vital for the future of the planet. In 1879 Thomas Edison patented the first filament lightbulb and cheap, safe lighting became available to most people. However, his design has only ever been about 2% efficient, meaning that for every 100J of electrical energy supplied to the bulb, only 2% is transferred into useful light, while 98% is wasted as thermal energy, heating the atmosphere. A modern white LED lightbulb is about 70% efficient, so this is a significant energy saving.

> **Efficiency** The ability of a device to transfer energy input into useful energy output. Usually expressed as a percentage, %.

Efficiency is defined as:

$$\text{efficiency} = \frac{\text{useful energy } (E) \text{ output (joules, J)}}{\text{total energy } (E) \text{ input (joules, J)}}$$

or, in terms of power,

$$\text{efficiency} = \frac{\text{useful power } (P) \text{ output (watts, W)}}{\text{total power } (P) \text{ input (watts, W)}}$$

where $P = \dfrac{E}{t}$.

It is also worth remembering that energy is not 'lost'. The law of conservation of energy says that:

total energy input = useful energy output + wasted energy output

Making links

You will need to be able to use power calculations when you come to look at the power of electrical devices on page 73 and mechanical devices on page 86.

Exam tip

Write down your full working in the exam because if you make a calculation error, such as pressing the wrong button on your calculator, you can still gain credit for correct working.

Maths skills

Unit prefixes or *standard form* are used to represent large or small quantities.

tera	giga	mega	kilo	centi	milli	micro	nano	pico
T	G	M	k	c	m	μ	n	p
10^{12}	10^{9}	10^{6}	10^{3}	10^{-2}	10^{-3}	10^{-6}	10^{-9}	10^{-12}

Worked example

A 5.0 kW electric winch pulls a car out of a ditch in 15 s. The useful energy (work done) on the car is 35 000 J.

a Calculate the total energy transferred as electricity into the winch.

b Calculate the percentage efficiency of the winch.

a Total energy supplied to the winch

$$E_t = P \times t = 5000 \text{ W} \times 15 \text{ s} = 75\,000 \text{ J}$$

b $\text{efficiency} = \dfrac{\text{useful energy output}}{\text{total energy input}} \times 100\% = \dfrac{35\,000 \text{ J}}{75\,000 \text{ J}} \times 100\% = 47\%$ (2 sf)

Practice question

1 The Drax biomass power station near Leeds burns wood pellets imported from the USA and Canada. The peak electrical output of the power station is 2.6 GW, and the power station burns the equivalent of 32 TJ of wood pellets per hour. Calculate the efficiency of the power station.

Maths skills

Significant figures (sf) refer to the number of numerals given in numerical data or answers. For example, if the power of a motor is written as 63.85 W, then this is to 4 sf. Written to 3 sf this is 63.9 W; to 2 sf, 64 W; and to 1 sf, 60 W. Standard form is the best way to write large and small numbers to the correct number of sf.

Worked example

An LED torch is 3.5% efficient. 7.9 mJ of useful light energy is emitted. Calculate the total electrical energy input to the torch.

$$\text{efficiency} = \frac{\text{useful energy output}}{\text{total energy input}} \times 100\%$$

$$\text{total energy input} = \frac{\text{useful energy output}}{\text{efficiency}} \times 100\%$$

$$= \frac{7.9 \times 10^{-3} \text{ J}}{3.5\%} \times 100\%$$

$$= 0.2257\ldots \text{ J}$$

Exam tip

As a general rule, you should give any calculated numerical answers to the same number of significant figures as the least precise piece of data used to calculate your answer. The exam paper will not tell you to write an answer to an appropriate number of significant figures – you will have to practise doing this frequently.

Written to 2 sf, the same sf as the data values, the answer is 0.23 J. Remember, zeros before the decimal point for numbers less than 1 are not significant, so the first significant number in this answer is the 2.

Practice questions

2   The electrical consumption of a digital watch battery is 21.6 J per year. The efficiency of the mechanism within the watch is 42%. Calculate the useful energy transferred per year as kinetic energy of the watch hands.

3   A small 15 W electric motor lifts a weight from the floor to a desk, transferring 24 J of energy in 4 s. Calculate the efficiency of the motor.

The efficiency of mechanical and electrical devices is hugely important

The efficiency of mechanical and electrical devices is hugely important. More efficient devices waste less energy, and this conserves the Earth's resources (such as crude oil and gas) and also improves the economics of the device, making it cheaper to run. More efficient domestic appliances, such as fridges, save individuals money, but large-scale efficient devices, such as the National Grid, save money for everyone. Improving the efficiency of electrical appliances means less electricity needs to be generated and so conserves resources and reduces pollution. Efficient devices and processes make the best use of available resources and waste less energy.

Efficiency can be improved by reducing wasted energy

Mechanical systems and machines

These primarily waste energy through thermal transfer, via friction, as moving parts move over each other. Friction can be reduced by *lubricants* such as oils, silicone sprays and sealed ball bearings.

Objects moving through fluids (such as the air or water) waste energy through *air resistance* or *drag*. This is a form of friction. Air resistance and drag can be reduced by making the objects more aerodynamic or hydrodynamic, by *streamlining* their shape.

Electrical or electronic devices

These devices waste energy through thermal transfer as well. *Electric current* flowing through components heats up the components due to their *resistance*. Reducing their resistance increases their efficiency. This can be achieved by using materials with lower resistivity, or by using some form of cooling, e.g. cooling fans or heat sinks. Large-scale computer data centres tend to be built in cold countries such as Iceland, as the ambient temperature helps to cool the electronics, making the systems more efficient. In some ac systems, the maximum efficiency is only obtained when the resistance of an input circuit is the same as that of an output circuit, for example, when the resistance of speakers and amplifiers is matched. In some cases, such as LED lighting, changing to a newer technology that is inherently more efficient drastically improves efficiency.

Thermal devices

Examples of these devices include heating systems, and energy here is wasted by *thermal transfer* to the surroundings. *Insulation* improves the efficiency of thermal systems, reducing thermal loss via conduction, convection and radiation. This is particularly important when designing houses to save money, save resources and reduce pollution.

Check your understanding and progress at **www.hoddereducation.co.uk/myrevisionnotes**

No device or process can be 100% efficient

Some energy will always be wasted through thermal transfer. Some LEDs and transformers can be over 95% efficient, but most large-scale heat engines such as power station furnaces and internal combustion engines in cars are never more than about 30% efficient.

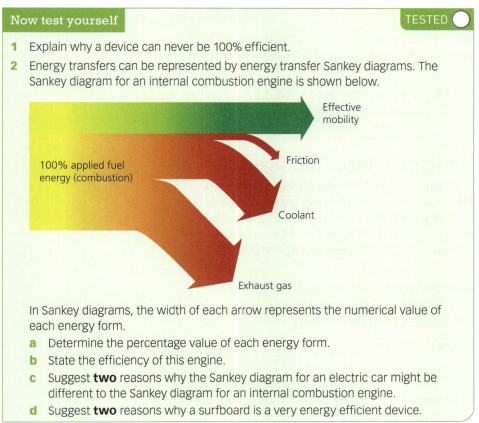
Maximising and minimising thermal transfer

Thermal transfer of heat via conduction, convection and radiation needs to be maximised in the examples below:

+ domestic and commercial heating systems
+ power stations involving the combustion of a fuel
+ passive solar heating systems
+ industrial processes that involve heating materials or components.

Once heat has been generated, for example by a domestic heating system, it is more efficient to reduce the thermal transfer wasted to the surroundings via conduction, convection and radiation. This can be achieved by using insulation systems. Examples of situations where thermal transfer needs to be minimised are:

+ domestic and commercial premises
+ industrial processes where hot fluids are pumped around a site, or where processes need to be kept at a particular temperature
+ hot/cold food distribution
+ cold weather clothing.

Conduction The transfer of thermal energy from hot to cold by the vibration of particles through a solid or a liquid.

Convection The transfer of thermal energy by the translation (movement) of particles from somewhere hot to somewhere cold.

Radiation The transfer of thermal energy by the emission of infrared electromagnetic radiation from hot objects to colder surroundings.

Now test yourself   TESTED

3 Explain why wood-burning stoves are usually made from steel and painted black.
4 Suggest why people are advised to wear several layers of clothes when the weather is cold.
5 Putting a lid on a take-away hot drink cup will reduce heat loss mostly by:
   A thermal conduction   C thermal radiation
   B thermal convection   D thermal evaporation

Buildings must be as energy efficient as possible

When buildings are designed, they must follow building regulations which require them to be as energy efficient as possible. The U-value (or thermal transmittance) of a structure such as a wall, floor, ceiling or window, is the amount of thermal energy flowing per second through $1\,m^2$ of a structure when there is a temperature difference of 1°C between the internal and external surfaces.

Good insulators have low U-values. Knowing the U-values of the different structures of a building, their area and the temperature difference allows an architect to calculate the thermal transfer from the building.

U-values can be calculated using the equation:

$$U = \frac{Q}{At\,\Delta T}$$

where Q is the thermal energy transferred (J); A is the area of the surface (m^2); t is the time (s); and ΔT is the temperature difference across the surface (°C).

Maths skills

The unit of U-value is a *compound unit*, made up of several other units. Units that would be divided (e.g. per m^2 or per °C) are written with a superscripted minus (-) sign. Joules per square metre per second per degree Celsius is written $J\,m^{-2}\,s^{-1}\,°C^{-1}$, although $J\,s^{-1}$ is equivalent to watts, W, so the unit is usually written $W\,m^{-2}\,°C^{-1}$.

Worked example

The uninsulated roof of a house has a total area of $52\,m^2$. Calculate the U-value of the roof, if 375W of thermal energy transfers through the roof when the temperature difference between the inside and the outside is 22°C.

$U = \dfrac{Q}{At\Delta T}$ but $P = \dfrac{Q}{t}$ so $U = \dfrac{P}{A\Delta T}$

Substitute in the numbers:

$$U = \frac{375\,W}{52\,m^2 \times 22\,°C} = 0.33\,W\,m^{-2}\,°C^{-1}\ (2\,\text{sf})$$

Practice questions

4 A wooden shed wall with an area of $6\,m^2$ transfers heat through it at a rate of 250W, when the temperature inside the shed is 20°C and the temperature outside is −5°C. Calculate the U-value of the shed wall.
5 Calculate the thermal energy that passes through an $8.4\,m^2$ brick wall with a U-value of $0.31\,W\,m^{-2}\,°C^{-1}$ in 6 hours, if the temperature outside the wall is 3°C and the temperature inside the wall is 19°C.

6 *U*-values are useful because most building structures are composite materials. A 6 m² wall consists of an internal layer of plasterboard, a layer of bricks, a layer of insulation, a further layer of bricks and then a layer of external render. When working out the total *U*-value of the wall, you just add all the *U*-values. Use the following data to determine the power transferred through the wall, when the temperature difference between the inside and the outside is 18 °C:

Material	*U*-value (W m^{-2} °C^{-1})
Plasterboard	1.8
Brick	2.2
Insulation	0.3
Render	1.4

Advantages and disadvantages of the different sources of useful energy

Table 1.10 Advantages and disadvantages of different energy sources

Energy source	Suitability	How it works	Advantages	Disadvantages
Fossil fuels (e.g. coal, oil, gas)	Large-scale production of electricity; domestic heating.	Fossil fuels are burned to release thermal energy for turning water into steam (power stations) OR for domestic hot water and heating systems.	Concentrated energy source. Reliable and predictable. Secure supply. Large scale. Easy to store.	Non-renewable energy source. Produces large quantities of CO_2. SO_2 and NO_x produced which contribute to acid rain. Other pollutants formed. Very inefficient. Eyesore.
Nuclear fuel	Large-scale production of electricity.	Nuclear fission of uranium fuel creates heat which can be used to produce steam for the generation of electricity.	Concentrated energy source. Reliable and predictable. Secure supply. Large scale. Easy to store. Does not produce large quantities of CO_2 or acid rain gases.	Non-renewable energy source. Potential for large-scale radioactive contamination (e.g. Chernobyl). Radioactive waste remains dangerous for thousands of years and needs secure long-term storage. Eyesore.
Solar power (heat)	Domestic or commercial hot water systems OR direct heating of south-facing buildings.	Passive solar power uses sunlight to heat water or air directly.	Renewable energy source. Free energy once installed.	Great for keeping suitably designed buildings warm. Solar-heated water is not very hot. Only works when it is sunny.

Energy source	Suitability	How it works	Advantages	Disadvantages
Solar power (light)	Generation of electrical energy. Solar power 'farms'; domestic and commercial solar panels for buildings. Low-power uses, e.g. calculators; chargers.	Light falls on a photovoltaic cell that generates a small voltage.	Renewable energy source. Free energy once installed. Works during daylight hours. Easy to install.	Does not work during the night. Only a small voltage is produced so larger panels needed for larger applications. Large areas are an eyesore.
Wind power	Large- and small-scale generation of electricity.	Wind turns a generator that transfers a store of kinetic energy into electricity.	Renewable energy source. Free energy once installed. Large wind farms can generate large amounts of electricity.	Only generates electricity when the wind blows over a threshold value – not as reliable and predictable as other sources. Only useful in windy areas. Some people think they are eyesores.
Wave power	Large-scale generation of electricity.	Moving water due to waves turns a generator that transfers a store of kinetic energy into electricity.	Renewable energy source. Free energy once installed. Potential to produce large quantities of electricity.	Only produces energy when there are waves. Not as reliable and predictable as other sources. Unproven technology and still 'experimental'.
Tidal power	Large-scale generation of electricity.	Moving water due to tidal currents turns a generator that transfers a store of kinetic energy into electricity.	Renewable energy source. Free energy once installed. Reliable and predictable. Large-scale generation possible.	Construction could destroy or alter estuary habitats. Only possible in limited locations. Requires a large tidal range.
Hydroelectric power	Large- or micro-scale generation of electricity.	Moving water stored behind a dam is channelled through pipes to generators that transfer a store of kinetic energy into electricity OR micro-scale generation in a river.	Renewable energy source. Free energy once installed. Reliable and predictable. Large-scale generation.	Construction of dam floods a river valley destroying wildlife habitats and sometimes human settlements. Only possible in areas with suitable river valleys.
Geothermal sources	Medium-scale generation of electricity.	Hot water from natural hot springs can be used to generate steam for production of electricity OR water for heating is pumped down to depth, heated and then pumped back to the surface.	Renewable energy source. Free energy once installed. Heating and hot water systems possible.	Only possible in areas with hot springs or 'hot rocks' at depth.

Check your understanding and progress at **www.hoddereducation.co.uk/myrevisionnotes**

Energy source	Suitability	How it works	Advantages	Disadvantages
Biomass	Large- and small-scale production of crops (e.g. wood) for burning in power stations OR for the production of biofuels.	Fast-growing wood is harvested and made into pellets which are burned in a power station OR crops are decomposed by microorganisms to produce biofuels.	Renewable energy source. Free energy once installed. Large- or small-scale production possible. Can be stored.	Burning wood creates CO_2. Large areas of land needed to grow crops.

Now test yourself

TESTED

6   Which of the following types of power station does not run on a renewable energy source?

   A Hydroelectric   **C** Wind

   B Nuclear   **D** Geothermal

7   About 1.8% of the UK's electricity is generated by hydroelectric power. Explain how a hydroelectric power station generates electricity.

8   What are the disadvantages of generating electricity from a hydroelectric power source?

Electricity and circuits

REVISED

Understanding how electrical circuits function is important to the design and operation of a huge range of electrical and electronic devices

These are the fundamental quantities that you need to know:

+ *Charge*, Q (coulombs, C) – this is the property of matter that causes electrical effects. Electrons are negatively charged and protons are positively charged. When we are dealing with electrical circuits, only some of the electrons are free to move, so these cause most of the electrical effects of a circuit. A single electron has a charge of only $1.602176634 \times 10^{-19}$ C, so it is difficult to measure the electrical effects of single electrons. Therefore, physicists study the overall effect of large numbers of electrons.

+ *Current*, I (amperes (amps), A) – the rate of flow of charge is called current. This is the amount of charge flowing, Q, in a given time, t. Current is measured by an ammeter connected in series with components. Charge, current and time are related by the equation:

$$I = \frac{Q}{t}$$

+ *Voltage*, V (volts, V) – this is a measure of the electrical energy supplied by the charge. Voltage is measured using a voltmeter. Voltmeters are always connected in parallel with components.

+ *Resistance*, R (ohms, Ω) – this is the opposition of a component to the flow of current through it. Conductors have low resistance and insulators have high resistance. Resistance can be measured directly using an ohmmeter or indirectly using a voltmeter–ammeter combination.

Voltage, current and resistance are related to each other by the equation (sometimes called Ohm's law):

$$I = \frac{V}{R}$$

+ *Power*, P (watts, W) – this is the rate at which electricity is transferred into other useful forms of energy by a device. Power is related to current and voltage by the equation:

$$P = IV$$

Making links

In Unit 2: Applied experimental techniques, you will do an experiment to determine the resistivity of a material (page 109). This requires you to use and apply the voltage, current and resistance equation.

Exam tip

You do not need to learn the equations as they are given to you on the Formulae sheet, but you do need to know how to rearrange them. You will need to practise doing this, by writing them out repeatedly, using cue cards or frequently answering questions where you need to do this.

Worked example

A circuit contains a 12V battery, an ammeter and a 9.0Ω resistor connected in series with each other.

a Calculate the current flowing in the circuit.

$$I = \frac{V}{R} = \frac{12\,\text{V}}{9.0\,\Omega} = 1.3333\,\text{A} = 1.3\,\text{A}\ (2\,\text{sf})$$

In questions like this, the second mark is frequently awarded only if the answer is given to the correct number of significant figures, which is two in this case.

b Calculate the power supplied to the resistor.

$$P = VI = 12\,\text{V} \times 1.3\,\text{A} = 15.6\,\text{W} = 16\,\text{W}\ (2\,\text{sf})$$

c Calculate the charge flowing through the resistor in 1.5 minutes.

$$I = \frac{Q}{t} \Rightarrow Q = It = 1.3\,\text{A} \times 90\,\text{s} = 117\,\text{C} = 120\,\text{C}\ (2\,\text{sf})$$

Now test yourself   TESTED

9 a A simple electrical circuit contains a 6.0V power supply, an ammeter and a 150Ω resistor all connected in series. Calculate the current I.

   b The current flows for 3.0 minutes. Calculate the charge Q that flows through the resistor.

   c Calculate the power of the resistor.

10 A torch circuit contains a battery, a switch and a filament lamp, which has a resistance of 1.8kΩ when a current of 2.5mA flows through it.

   a Draw a circuit diagram for this torch.

   b Calculate the voltage of the battery.

   c Calculate the power of the lamp.

11 A 12V cordless drill draws a current of 12.5A drilling a hole.

   a Calculate the resistance of the drill during its operation.

   b Calculate the power of the drill.

Current in a wire makes the wire heat up

A power supply such as a battery or a power supply unit provides the voltage (energy) to move the electric charges (electrons) through the wire. All the charges start to move at once (like the links in the chain on a bicycle). As the electrons flow through the wire, the structure of the wire gets in the way and collisions occur between the electrons and the structure of the wire (and between the electrons themselves). This causes the transfer of thermal energy to the structure of the wire so the wire starts to heat up. The greater the current and the greater the resistance, the greater the rate of heat loss. This is given by the equation:

rate of heat loss = I^2R

Now test yourself   TESTED

12 An electric fire element has a resistance of 8.9Ω when a current of 13A flows through it.

   a Calculate the voltage across the element.

   b Calculate the rate of heat loss.

13 The element of a hairdryer loses heat at a rate of 1.9kW, and has a resistance of 150Ω.

   a Calculate the current flowing through the element.

   b Calculate the voltage across the element.

Series and parallel circuits

In **series circuits** such as the one shown in Figure 1.42, the current is the same at every point around the circuit.

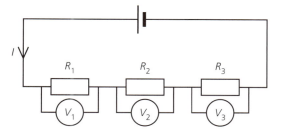

Figure 1.42 Circuit diagram showing resistor combinations in series

The current I is the same through each of the resistors R_1, R_2 and R_3.

This means that it does not matter where you measure the current in a series circuit – the value will be the same.

The voltage in a series circuit is shared between the components in series. The sum of all the voltages into the circuit (supplied by batteries, power supplies, etc.) is equal to the sum of the voltages out of the circuit (across each component). In the case of Figure 1.42: $V_{in} = V_1 + V_2 + V_3$.

The total resistance of components in series is the sum of all the resistances. In Figure 1.42: $R_{total} = R_1 + R_2 + R_3$.

In **parallel circuits** such as the one shown in Figure 1.43, current splits or recombines at junctions.

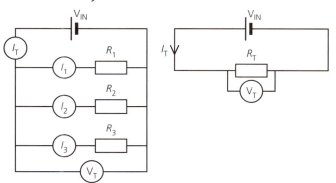

Figure 1.43 Circuit diagram showing resistors connected in parallel

The current out of the power supply, I_T, is the same as the current returning to the power supply, but at the junctions: $I_T = I_1 + I_2 + I_3$.

As the resistors are connected in parallel, the voltage, V_T, across each resistor is the same as the voltage, V_T, of the power supply.

Calculating resistance in parallel is more complex. Adding a resistance in parallel makes the overall resistance less. The equation for calculating resistance in parallel is:

$$\frac{1}{R_{Total}} = \frac{1}{R_1} + \frac{1}{R_2} + \frac{1}{R_3}$$

> ### Worked example
>
> A circuit contains a 400 Ω resistor connected in parallel with a 600 Ω resistor. Calculate the total resistance of this circuit.
>
> $$\frac{1}{R_{Total}} = \frac{1}{R_1} + \frac{1}{R_2} = \frac{1}{400\,\Omega} + \frac{1}{600\,\Omega} = \frac{1}{240\,\Omega} \Rightarrow R_{Total} = 240\,\Omega$$

> **Series circuit** A circuit where the components are connected in a complete loop, one after another.
>
> **Parallel circuit** A circuit where two or more components are connected to the same points in the circuit with junctions.

> **Exam tip**
>
> When calculating resistances in circuits with series **and** parallel combinations of components, calculate the parallel parts first.

My Revision Notes: AQA Applied Science Suitable for Level 3 and Level 3 Extended Certificates

14 A circuit contains three 75 Ω resistors connected in series. The resistors are connected in series with a 4.5 V battery.

   a Calculate the total resistance of the circuit.

   b Calculate the current flowing through each resistor.

   c Calculate the heat lost by each resistor.

15 A lamp rated as 12 V 0.80 mA is connected in parallel with a 12 kΩ resistor and an 8.0 kΩ resistor. All three are connected in parallel with a 12 V power supply.

   a Draw the circuit.

   b Calculate the current drawn by the 12 kΩ resistor and the 8.0 kΩ resistor.

   c Calculate the total current drawn from the 12 V power supply.

   d Calculate the total resistance of the circuit.

Potential dividers control voltages and can act as sensors

Potential dividers involve the use of two components with resistance. Consider the circuit in Figure 1.44.

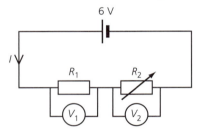

Figure 1.44 Circuit diagram of a potential divider

If the fixed resistor, R_1, is 1000 Ω and the variable resistor, R_2, is set to 1000 Ω, the resistance of each resistor is the same, so the voltage across each one is the same, and the total voltage, 6 V, is shared equally between the two resistors (3 V and 3 V). Decreasing the resistance of the variable resistor decreases the voltage across it, so the voltage across the fixed resistor rises. This circuit provides a mechanism for *varying voltage*, for example, in a variable power supply. Connecting the variable resistor to a computer allows for automatic and remote control.

The second use of a potential divider is as a sensor circuit. Consider Figure 1.45.

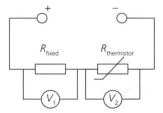

Figure 1.45 A potential divider as a sensor circuit

The resistance of the thermistor, $R_{thermistor}$, varies with temperature. Increasing temperature causes the resistance of the thermistor to decrease. This causes the voltage V_2 to drop as the temperature increases, so V_1 increases. The potential divider is now acting as a temperature sensor. This has a huge number of applications as any component whose resistance changes with a physical quantity (such as light intensity; sound intensity; strain; magnetic field) can be used to turn a circuit into an electric/electronic sensor. Connecting the voltage across the fixed resistor, R_{fixed}, to a computer allows for automatic and remote sensing.

16 An electronics engineer is designing a circuit for a portable charger. The main battery is 10.8 V, but the charger needs to be used to charge a phone with a voltage of 4.7 V. The engineer has already incorporated a 3.9 kΩ fixed standard resistor into the circuit. What resistor should the engineer use in a potential divider circuit with the 3.9 kΩ resistor to provide a voltage of 4.7 V, with maximum current drawn from the main battery?

17 A portable variable power supply is designed with a 220 Ω fixed standard resistor connected in series with a 75–550 Ω variable resistor in a potential divider circuit powered by an 18 V battery. Calculate the minimum and maximum voltages that can be taken from a connection across the variable resistor.

18 An electronic temperature sensor for an oven uses a 5.5 kΩ fixed standard resistor in series with a thermistor that is mounted inside the wall of the oven. The sensor potential divider circuit is powered by a 6.5 V rechargeable battery. The figure shows how the resistance of the thermistor varies with temperature.

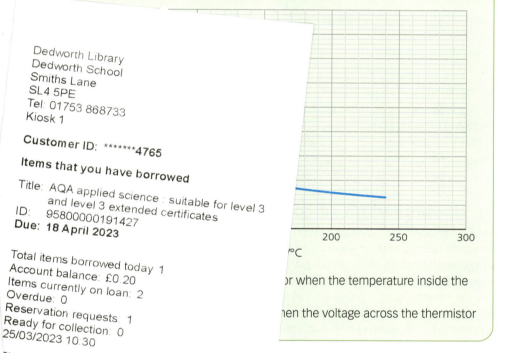

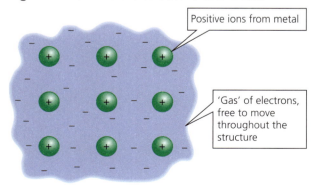

°C

...or when the temperature inside the

...hen the voltage across the thermistor

...ectricity involves movement of electrons

Metals are excellent conductors of electricity, because their structure contains lots of electrons that are free to move throughout the structure, behaving like a 'gas' of electrons. A *free* electron is an electron that has been freed from the outer shell of an atom when the atoms come together to form the solid metal. Figure 1.46 shows the structure of a metal.

Positive ions from metal

'Gas' of electrons, free to move throughout the structure

Figure 1.46 The structure of a metal

The free electrons are moving constantly in random directions with a mean high velocity that depends on the temperature. When a voltage is applied across the ends of the metal conductor, the 'gas' of electrons gradually drifts towards the positive voltage. The conductivity of the metal conductor depends on the number of free electrons and the arrangement of the positive ion cores of the rest of the structure. Increasing the temperature of the metal increases the mean velocity of the free electrons, which increases the rate of collisions between the electrons and the positive ion cores, so increasing the resistance.

Semiconductors do not conduct electricity as well as metals do. Their structure contains fewer electrons that are free to move throughout the structure. Semiconducting components can be manufactured with a range of different structures, allowing for the design and control of their conductivity and electrical properties. Many semiconductors are negative temperature coefficient (ntc) materials. Increasing their temperature causes more conducting electrons to be freed from the structure, increasing their conductance and decreasing their resistance.

Thermistors and light-dependent resistors are examples of semiconducting components.

Now test yourself	TESTED ⬤

19 Explain why copper is such a good conductor of electricity.

20 Semiconductors can be 'doped' by the addition of atoms of elements such as phosphorous into their structure to add extra free electrons into the structure. What is the effect on the conductivity of a semiconductor of adding phosphorous atoms?

21 Why does the resistance of a metal increase as the metal is heated?

Making links

In Unit 2: Applied experimental techniques, you will do an experiment to determine the resistivity of a material. This requires you to think about the electrical structural properties of different materials.

VI graphs tell us about the behaviour of components

Voltage, V, against current, I, graphs are sometimes called electrical characteristics. They comprehensively describe the behaviour of electrical and electronic components. The voltage is plotted on the y-axis and the current is plotted on the x-axis. The voltage divided by the current, at any point on the graph, gives the resistance of the component. Therefore, calculating the *gradient* (or slope) of the VI graph at any point will give the resistance of the component. High resistance is represented by steep lines and low resistance by shallow lines. VI graphs are sometimes plotted showing negative voltages and currents. This shows how the component behaves when it is connected in the opposite direction to normal (called reverse bias). Standard resistors have constant resistance, so their VI graph is a straight line. A steeper line represents higher resistance.

Figure 1.47 shows the VI graph for two fixed standard resistors.

Electrical characteristic
A voltage–current (VI) graph, with current plotted on the *x*-axis and voltage plotted on the *y*-axis.

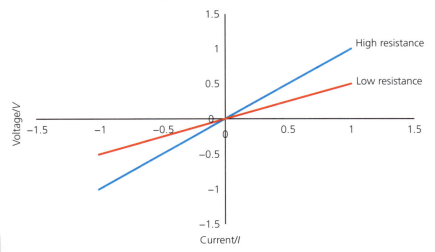

Figure 1.47 A VI graph for two fixed standard resistors

Check your understanding and progress at **www.hoddereducation.co.uk/myrevisionnotes**

Figure 1.48 shows the VI graphs for a filament lamp, a thermistor and an LDR.

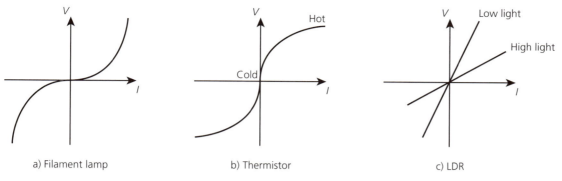

a) Filament lamp   b) Thermistor   c) LDR

Figure 1.48 VI graphs for (a) a filament lamp, (b) a thermistor and (c) a light-dependent resistor

In Figure 1.48a, the resistance of the metal filament in the lamp increases with current because the metal is heating up. The gradient of the line gets steeper with increasing current.

In Figure 1.48b, the resistance of the thermistor decreases with increasing current, because more free electrons are released by the structure as it heats up due to increasing current.

In Figure 1.48c, the resistance of the LDR is higher in low light intensities, as fewer free electrons are released by the structure, and vice versa in high light intensities.

Now test yourself   TESTED ◯

22 A circuit contains two standard resistors, R_1 and R_2, and a lamp. The VI graph for all three components is shown below.

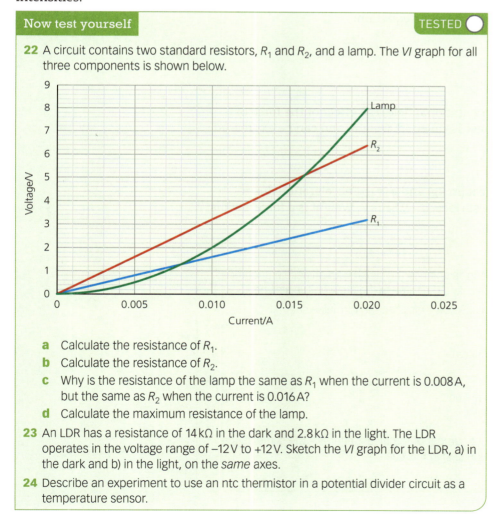

a Calculate the resistance of R_1.
b Calculate the resistance of R_2.
c Why is the resistance of the lamp the same as R_1 when the current is 0.008 A, but the same as R_2 when the current is 0.016 A?
d Calculate the maximum resistance of the lamp.

23 An LDR has a resistance of 14 kΩ in the dark and 2.8 kΩ in the light. The LDR operates in the voltage range of –12 V to +12 V. Sketch the VI graph for the LDR, a) in the dark and b) in the light, on the *same* axes.

24 Describe an experiment to use an ntc thermistor in a potential divider circuit as a temperature sensor.

How do I apply Newton's Laws of Motion?

Newton's Laws of Motion are used to predict the behaviour of moving objects. They can be used to predict motion and the way that objects interact with each other.

Newton's *First Law of Motion* states that an object will stay at rest or continue moving at a constant velocity until it is acted on by a resultant force.

Newton's First Law introduces the concept of inertia, which is the opposition of an object to having its motion altered. Objects that have a large amount of inertia tend to be massive, or moving very fast, or both. If an object has high inertia, it is difficult to start the object moving, difficult to change the direction of its motion, and difficult to stop it moving.

> **Inertia** The property of a moving object that opposes a change in its motion. Inertia is linked to momentum.

Objects that are stationary must have zero resultant force, so any forces acting on the object must be balanced. Similarly, an object that is moving will continue to move at constant velocity, unless acted upon by a resultant force. Objects subject to unbalanced or resultant forces will accelerate (or decelerate).

Newton's *Second Law of Motion* describes the motion of an object when it is subject to a resultant force. Although Newton formally stated the law in terms of rate of change of momentum, it is best described as the relationship between the resultant force, F (in N), mass, m (in kg) and acceleration, a (in m s^{-2}), of the object.

Newton's second law is represented by the equation:

force = mass × acceleration

or

$F = ma$

For objects inside the Earth's gravitational field, the effect of the force of gravity acting on an object's mass is called its weight, and can be expressed by the equation:

weight = mg

where m is the mass of the object (in kg) and g is the acceleration due to gravity (in m s^{-2}). At the Earth's surface, $g = 9.8$ m s^{-2}. Weight is a force, so it is measured in newtons, N.

Worked example

The two-person crew of a 171 kg bobsleigh push the sleigh with an acceleration of 2.25 m s^{-2} at the start of a race. Calculate the force of the crew on the sleigh.

1   Write down the equation that you know (this is on the formulae sheet):

  $F = ma$

2   Insert the data given in the question.

  $F = 171 \, \text{kg} \times 2.25 \, \text{m s}^{-2}$

3   Calculate the answer and add the unit.

  384.75 N

4   Write your answer to the correct number of significant figures – the same as the least precise data in the question (3 sf in this case).

  385 N

Newton's *Third Law of Motion* states that forces always occur in pairs, called action and reaction forces. When object A exerts an action force on object B, then object B exerts an equal and opposite reaction force on object A. Although the two forces are equal and opposite, they act on different objects, and the forces in the pair are always of the same type (direct contact, e.g. tension or friction; or action-at-a-distance, such as gravity or magnetism).

Now test yourself

TESTED

25 Why does a large truck travelling at $20\,m\,s^{-1}$ have more inertia than a motorbike travelling at the same velocity?

26 Calculate the resultant force acting on a 54 kg sprinter accelerating out of their blocks at $3.2\,m\,s^{-2}$.

27 Explain in terms of Newton's Third Law of Motion why a bungee jumper remains stationary at the end of a bungee cord as she is hanging waiting to be let down.

Momentum and Newton's Third Law of Motion

Momentum, symbol p, is a measure of the inertia of an object. Momentum is defined as: momentum = mass × velocity or $p = mv$.

The greater the inertia of an object, the greater its momentum. Momentum has the compound unit $kg\,m\,s^{-1}$ and it is a vector quantity (like velocity and force), so it has magnitude and direction and can be represented by an arrow on diagrams. Like energy, momentum is always conserved during the interaction of objects. This is called the Law of Conservation of Momentum, which can be written as:

total momentum before interaction = total momentum after interaction

If two objects collide, the loss of momentum of one object must be balanced by the gain of momentum by the second object. This is a consequence of Newton's Third Law of Motion and is usefully applied in a wide range of applications from rockets to sports science. For example, in tennis, the momentum lost by the racquet is gained by the ball. Because the ball has less mass than the racquet, the ball moves away from the racquet at a much higher speed than the approach speed of the racquet.

Worked example

During a tennis serve, a 310 g racket hits a (momentarily) stationary 58 g tennis ball, with a velocity of $6.7\,m\,s^{-1}$, and then stops moving. Calculate the velocity of the ball immediately after the serve.

1 Using the Law of Conservation of Momentum:

total momentum before the collision = total momentum after the collision

2 Write out the equation:

$$\left(mu\right)_{ball} + \left(MU\right)_{racket} = \left(mv\right)_{ball} + \left(MV\right)_{racket}$$

3 The initial velocity of the ball, u, and the final velocity of the racket, V, are both zero. So:

$$\left(MU\right)_{racket} = \left(mv\right)_{ball}$$

4 Rearranging to make the velocity of the ball the subject and substituting the numbers:

$$v_{ball} = \frac{\left(MU\right)_{racket}}{m_{ball}}$$

$$= \frac{310 \times 10^{-3}\,kg \times 6.7\,m\,s^{-1}}{58 \times 10^{-3}\,kg}$$

$$= 35.810\,m\,s^{-1} = 36\,m\,s^{-1}\,(2\,sf)$$

Newton's Second Law of Motion can be defined in terms of momentum. If the initial velocity of an object is u (in $m\,s^{-1}$), its final velocity is v (in $m\,s^{-1}$) and the time for the change is t (in s), then the acceleration, a (in $m\,s^{-2}$) is given by:

$$a = \frac{v-u}{t}$$

So, $F = ma$ can be re-written as:

$$F = m\left(\frac{v-u}{t}\right) = \frac{mv-mu}{t} = \frac{p_{\text{final}} - p_{\text{inital}}}{t} = \frac{\Delta p}{t}$$

In other words, force is the rate of change of momentum. Two common applications of this equation involve automotive engineering and sports science. Cars are designed with front and back 'crumple zones' that 'crumple' during collisions. This takes time and so increases the time of the collision. Increasing the collision time reduces the rate of change of momentum and hence decreases the force of the collision, making it safer for the occupants of the car. A similar effect is used in sports science when the sport involves any form of collision. For example, boxers, martial arts fighters and rugby players frequently wear body padding to reduce impact forces. The padding is designed to deform during an impact, which increases the collision time and decreases the impact force.

Worked example

During a car crash, a 950 kg car travelling at $26\,m\,s^{-1}$, collides with a stationary bollard. The velocity of the car is reduced to 0 in 0.21 s. A similar car, travelling at the same velocity, also collides with a stationary bollard. This car is fitted with a crumple zone, which increases the collision time to 0.55 s. Calculate the difference between the force of the impacts, with and without the crumple zone.

1   Write down the equation that you know and substitute the data for the first vehicle:

$$F = \frac{mv - mu}{t}$$
$$= \frac{950\,kg \times 0\,m\,s^{-1} - 950\,kg \times 26\,m\,s^{-1}}{0.21\,s}$$
$$= 117\,619\,N = 120\,kN\ (2\ sf)$$

The negative sign produced by this equation implies that the force acts in the opposite direction to the motion. Only the magnitude is shown here.

2   Repeat for the second vehicle:

$$F = \frac{mv - mu}{t}$$
$$= \frac{950\,kg \times 0\,m\,s^{-1} - 950\,kg \times 26\,m\,s^{-1}}{0.55\,s}$$
$$= 44\,909\,N = 45\,kN\ (2\ sf)$$

3   The difference between the two forces:

Difference = 120 kN − 45 kN = 75 kN

Now test yourself

TESTED

28 An air gun pellet of mass 6.7 g is fired out of an air gun at a velocity of 250 $m\,s^{-1}$. Calculate the momentum of the pellet.

29 During a serve in table tennis, a 0.15 kg bat travelling at 0.75 $m\,s^{-1}$ hits a stationary table tennis ball of mass 2.7 g, and the bat stops moving. Calculate the velocity of the ball immediately after the impact.

30 The seat belts of a car are designed to stretch during an impact and must be replaced after a crash. Explain how this reduces the force of the impact on the driver and passengers.

Check your understanding and progress at **www.hoddereducation.co.uk/myrevisionnotes**

Newton also formalised equations that can be used to calculate the motion of objects

In these equations, symbols are used to describe the motion of an object (sometimes called the suvat symbols).

The equations and symbols are:

$(\text{average})\ v = \dfrac{s}{t}$ s = distance moved (in m)

$v = u + at$ u = initial velocity (in m s⁻¹)

        v = final velocity (in m s⁻¹)

$v^2 = u^2 + 2as$ a = acceleration (in m s⁻²)

$s = ut + \dfrac{1}{2}at^2$ t = time taken (in s)

> ## Exam tip
>
> When answering questions involving the equations of motion, it is best to write the acronym suvat vertically down the page and then use the question to write the given data values next to the symbols with a question mark, ?, next to the quantity you are asked to calculate. This will make it easier to identify which equation of motion you need to use to answer the question.

> ## Worked example
>
> A skier is travelling at a constant velocity of 16 m s⁻¹ before accelerating at 1.2 m s⁻² for 3.5 s. Calculate the distance travelled by the skier.
>
> 1 Write down the suvat acronym, with the data from the question:
>
> $s = ?$
>
> $u = 16\,\text{m s}^{-1}$
>
> $v = -$
>
> $a = 1.2\,\text{m s}^{-2}$
>
> $t = 3.5\,\text{s}$
>
> 2 Then write down the equation needed:
>
> $s = ut + \dfrac{1}{2}at^2$
>
> 3 Substitute the data into the equation:
>
> $s = \left(16\,\text{m s}^{-1} \times 3.5\,\text{s}\right) + \left(\dfrac{1}{2} \times 1.2\,\text{m s}^{-2} \times \left(3.5\,\text{s}\right)^2\right)$
>
> 4 Calculate the answer, add the unit and write to the correct number of significant figures (2).
>
> $s = 63.35\,\text{m} = 63\,\text{m}\ \left(2\ \text{sf}\right)$

> ## Now test yourself   TESTED ◯
>
> 31 An athlete is running at 9.5 m s⁻¹ and then accelerates towards the finish line at 1.2 m s⁻² for 2.5 s. Calculate the athlete's velocity as they cross the finish line.
>
> 32 A track cyclist is initially travelling at 12 m s⁻¹, but then accelerates at 0.85 m s⁻² for the last 50 m of the race. Calculate the velocity of the cyclist as they cross the finish line.
>
> 33 A bungee jumper, initially at rest, jumps off a bridge attached to a bungee and falls for 2.7 s. Calculate the distance they fall in this time. Assume that there is no air resistance and $g = 9.8\,\text{m s}^{-2}$.

There are two types of motion graph

The first is a displacement–time graph (s–t) and the second is a velocity–time graph (v–t).

> **Worked example**
>
> Figure 1.49 shows a displacement–time graph for a football. Describe in words the motion of the football and calculate any values.
>
>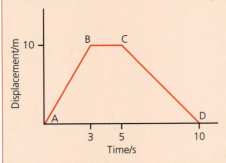
>
> **Figure 1.49** A displacement–time graph for a football
>
> 1 You need to describe each section of the motion:
>   The football is stationary at the origin.
>
> 2 Calculate values from the graph:
>   It is then kicked and moves away from the origin for 3 seconds. To calculate the velocity:
>
> $$\text{velocity} = \frac{\text{vertical interval}}{\text{horizontal interval}} = \frac{10\,\text{m}}{3\,\text{s}} = 3.3\,\text{m}\,\text{s}^{-1}\ (2\ \text{sf})$$
>
> 3 Continue to describe the motion:
>   The ball is trapped and kept stationary for a further 2 seconds, 10 m away from the origin.
>
> 4 Final calculation:
>   The ball is kicked back to the origin for 5 seconds. To calculate the velocity:
>
> $$\text{velocity} = \frac{\text{vertical interval}}{\text{horizontal interval}} = \frac{10\,\text{m}}{5\,\text{s}} = 2\,\text{m}\,\text{s}^{-1}$$

+ Stationary objects have a horizontal line on an s–t graph.
+ Objects moving at constant velocity have straight, sloping lines. Positive gradients show motion away from the origin and negative gradients show motion towards the origin.
+ The gradient of the line is the velocity.
+ Curved lines indicate acceleration or deceleration.

Gradient The slope of a line on a graph (vertical interval ÷ horizontal interval).

Displacement Of an object, is its distance away from an origin in a given direction.

Worked example

Figure 1.50 shows the velocity–time graph for a sprinter in a race. Describe the motion of the sprinter, calculating any values.

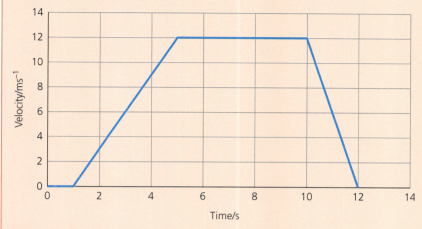

Figure 1.50 A velocity–time graph for a sprinter in a race

1 You need to describe each section of the motion:
The sprinter is stationary in her blocks for 1 second.

2 Calculate acceleration:
She moves away from the blocks with a constant acceleration for 4 seconds. To calculate the acceleration:

$$\text{acceleration} = \frac{\text{vertical interval}}{\text{horizontal interval}} = \frac{12\,\text{m s}^{-1}}{4\,\text{s}} = 3\,\text{m s}^{-2}$$

3 Continue to describe the motion:
The sprinter then runs at a constant velocity of 12 m s^{-1} for 5 seconds.

4 Calculate the deceleration:
The sprinter decelerates for 2 s. To calculate the deceleration:

$$\text{deceleration} = \frac{\text{vertical interval}}{\text{horizontal interval}} = \frac{12\,\text{m s}^{-1}}{2\,\text{s}} = 6\,\text{m s}^{-2}$$

+ Stationary objects on a v–t graph have a horizontal line along the time axis.
+ Objects moving at constant velocity have horizontal lines above the time axis.
+ Objects undergoing constant acceleration have straight sloping lines with a positive gradient, and objects undergoing constant deceleration have straight sloping lines with a negative gradient.
+ The total distance travelled by the object is equal to the *area* under the v–t graph.

My Revision Notes: AQA Applied Science Suitable for Level 3 and Level 3 Extended Certificates

34 A skateboarder performs a series of moves on their board. A displacement–time graph of their motion is shown in the figure.

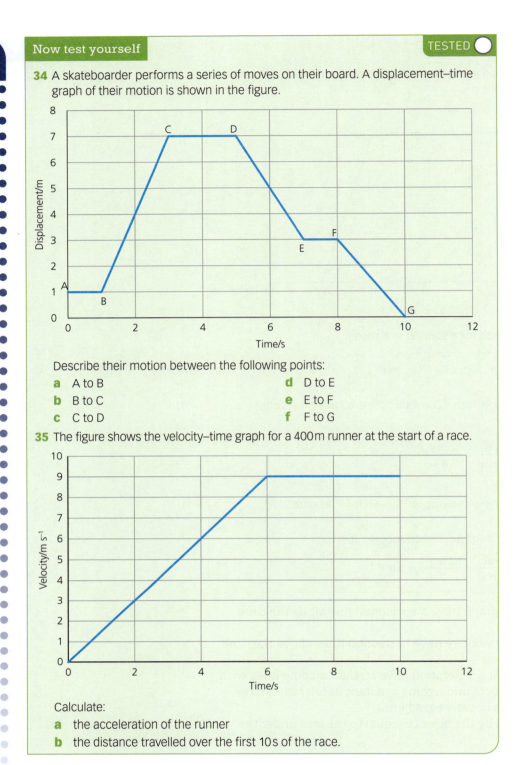

Describe their motion between the following points:

a A to B   **d** D to E

b B to C   **e** E to F

c C to D   **f** F to G

35 The figure shows the velocity–time graph for a 400m runner at the start of a race.

Calculate:

a the acceleration of the runner

b the distance travelled over the first 10s of the race.

We can calculate the energy and power of motion

Moving objects possess a store of kinetic energy (KE). This depends on the mass of the moving object, m (in kg), and its velocity, v (in $m\,s^{-1}$). The equation for kinetic energy is:

$$KE = \frac{1}{2}mv^2$$

Maths skills

Questions involving the properties and quantities used to describe moving objects are frequently used to test a candidate's knowledge of how to convert units. Questions about motion often involve larger prefixes, such as kilo, k (e.g. kilometres) or mega, M (e.g. megawatts). Remember, k means 'thousand' or 10^3, and M means 'million' or 10^6.

Check your understanding and progress at **www.hoddereducation.co.uk/myrevisionnotes**

Objects that are moving up and down (inside the Earth's gravitational field) gain or lose a store of gravitational potential energy (GPE). GPE depends on: the mass of the object, m (in kg); the gravitational field strength, g (in $N kg^{-1}$); and the height interval, h (in m). The equation for GPE is:

$$GPE = mgh$$

Worked example

A 5.2 g acorn falls from a tree from a height of 3.4 m. Calculate the velocity at which the acorn hits the floor. The acceleration due to gravity, $g = 9.8\,N kg^{-1}$.

1   Calculate the GPE of the acorn in the tree:

$$GPE = mgh$$
$$= 5.2 \times 10^{-3}\,kg \times 9.8\,N\,kg^{-1} \times 3.4\,m$$
$$= 0.173264\,J = 0.17\,J\ (2\,sf)$$

2   Equate the GPE with the KE:

$$GPE = KE = 0.17\,J = \frac{1}{2}mv^2$$

3   Rearrange the equation to make v the subject:

$$v = \sqrt{\frac{2 \times KE}{m}} = \sqrt{\frac{2 \times 0.17\,J}{5.2 \times 10^{-3}\,kg}}$$
$$= 8.086\,m\,s^{-1} = 8.1\,m\,s^{-1}\ (2\,sf)$$

The mechanical power, P (in watts, W) of a system is defined as the amount of energy transferred, E (in joules, J) per unit time, t (in seconds, s). The equation for the mechanical power of a system is:

$$P = \frac{E}{t}$$

Practice questions

7   Calculate the kinetic energy of a 1.8×10^3 kg van travelling at $26\,m\,s^{-1}$.

8   A football of mass 430 g is thrown vertically into the air through a distance of 14 m by a goalkeeper. Calculate the increase in GPE of the ball at the top of the throw. The gravitational field strength, $g = 9.8\,N kg^{-1}$.

9   A hairdryer is rated with a power of 1500 W. Calculate the total energy transferred by the hairdryer in 5 minutes drying someone's hair.

Summary

+ The efficiency of a device is important; it can be improved but can never be 100% efficient.

$$efficiency = \frac{useful\ energy\ (or\ power)\ output}{total\ energy\ (or\ power)\ input}$$

+ In some situations, thermal transfer needs to be maximised and in others, minimised.
+ U-values allow the comparison of the insulating properties of materials, where: $U = QAt\Delta T$.
+ Useful energy is generated using a range of different non-renewable and renewable sources: fossil fuels; nuclear fuels; solar power; wind power; wave power; tidal power; hydroelectric power; geothermal sources; and biomass. Each one has advantages and disadvantages.
+ Experiments can be carried out to measure efficiency.
+ It is possible to calculate: current; voltage; power; resistance and rate of heat loss in a range of series and parallel electrical circuits, using the formulae: $I = \frac{Q}{t}$; $P = IV$; $I = \frac{V}{R}$; rate of heat loss I^2R.
+ The total resistance of series and parallel resistors is given by:

series: $R_{total} = R_1 + R_2 + R_3$

parallel: $\frac{1}{R_{total}} = \frac{1}{R_1} + \frac{1}{R_2} + \frac{1}{R_3}$

+ Potential dividers can be used to control voltages or in sensor circuits.
+ Conductors like metals and semiconductors conduct electricity because they have free electrons within their structure and their resistance varies with temperature.

+ Thermistors vary their resistance with temperature and light-dependent resistors (LDRs) vary their resistance with light intensity.
+ The *VI* graphs of standard resistors, thermistors and lamps have different shapes and can be used to find resistance.
+ Newton's First Law of Motion applies to both stationary and moving objects and describes inertia.
+ Newton's Second Law of Motion leads to the formulae: $F = ma$ and $\text{weight} = mg$.
+ Newton's Third Law of Motion leads to: the concept of momentum $p = mv$; the Law of Conservation of Momentum; and the definition of force, $F = \dfrac{\Delta p}{t}$.

+ Newton's equations of motion can be used to analyse the motion of objects, where: $(\text{average})\ \text{speed} = \dfrac{s}{t}$; $v = u + at$; $v^2 = u^2 + 2as$; $s = ut + \dfrac{1}{2}at^2$.
+ Motion can be described by graphs of: displacement against time; and velocity against time.
+ The gravitational potential energy of an object, GPE, is given by: $\text{GPE} = mgh$.
+ The kinetic energy of a moving object, KE, is given by: $\text{KE} = \dfrac{1}{2}mv^2$.
+ The power of a mechanical system is given by: $P = \dfrac{E}{t}$.

Exam practice

1 The figure below shows a transformer circuit for a model bathroom razor socket. These sockets use transformers to isolate the mains supply so that the risk of electrocution is eliminated.

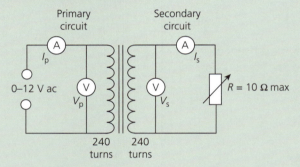

This circuit is used by a student to investigate the efficiency of the transformer, and how this changes with the resistance of the 10 Ω variable load resistor connected in series with the secondary circuit.

The electrical power in the primary and secondary circuits is given by:

$$P_p = V_p \times I_p \text{ and } P_s = V_s \times I_s$$

Describe how the student would do this experiment. In your description you should state:
+ the variables involved with this experiment
+ the measurements they should take
+ how they should ensure that their results are valid. [5]

2 The student's data is shown in the table below.
2.1 Complete the table and show your working. [3]
2.2 On graph paper, plot a graph of efficiency against resistance and add a suitable line of best fit. [3]

2.3 Describe the pattern shown by your graph. [2]
2.4 Use your graph to estimate the resistance when the efficiency is at a maximum. [1]
2.5 The resistor is replaced with one of two small motors, A and B. Motor A has a resistance of 4 Ω and Motor B has a resistance of 8 Ω.
State and explain why Motor A is a better choice for the motor in a shaving razor powered from this transformer. [2]
2.6 In a battery-operated razor, the power supplied to the motor is 2.25 W. The motor is only 38% efficient. Calculate the useful power converted into kinetic energy by the motor. [2]
2.7 Give **one** reason why an electric motor is never 100% efficient. [2]

3 Formula E racing is a motor racing championship for electric cars.

The cars have many features common to Formula 1 cars, but the engine is electric, there is no transmission (gears) system or fuel tank, and instead there are heavy batteries.
3.1 Suggest and explain how the following features increase the efficiency of a Formula E car: [4]
3.1.1 shape
3.1.2 friction between the moving parts in the motor.

Resistance, R/Ω	Primary circuit		Secondary circuit		Efficiency/%
	Potential difference, V_P/V	Current, I_P/A	Potential difference, V_S/V	Current, I_S/A	
2	12.0	0.54	11.9	0.50	91
4	12.0	0.54	11.8	0.54	
6	12.0	0.54	11.8	0.53	96
8	12.0	0.54	11.9	0.51	
10	12.0	0.54	11.9	0.50	92

Check your understanding and progress at **www.hoddereducation.co.uk/myrevisionnotes**

The table below summarises some key features of Formula E and Formula 1 cars:

Feature	Formula E car	Formula 1 car
Races	Street circuits	Mostly racetracks
Mass, kg	898	702
Maximum output power, kW	200	710
Acceleration, $m\,s^{-2}$ (0–100 km h^{-1})	9.3	13.2
Top speed, $m\,s^{-1}$	63	105
Maximum noise level, dB	80	134
Engine efficiency, %	75	50

3.2 Use the data in the table to calculate the total input power to each engine. Show your working. [3]

    3.2.1 Formula E engine

    3.2.2 Formula 1 engine.

3.3 Suggest a reason why the Formula E engine is more efficient than the Formula 1 engine. [2]

3.4 Suggest a reason why the maximum power of a Formula E racing car is restricted to 200 kW with a maximum speed of 63 m s^{-1}. [1]

3.5 Explain why reducing the mass of the batteries (but keeping the battery energy capacity the same) in a Formula E car will increase the efficiency of the engine. [2]

4 Pizzas are usually delivered warm to your door from the nearest pizza parlour. The pizzas are kept warm by a combination of a stiff cardboard box and an insulated pizza box carrier, which has a silvered material on its inside surface, a bubble-wrap insulator surrounding and a close-fitting Velcro sealed opening flap.

4.1 Describe how the following design features reduce thermal loss from the pizza. You should state which thermal transfer process is being minimised in each case. [4]

    4.1.1 Stiff cardboard box

    4.1.2 Silvered inside surface

    4.1.3 Bubble-wrap insulator

    4.1.4 Close-fitting Velcro sealed opening flap

4.2 The pizza carrying system has a surface area of 0.58 m² and a total U-value of 0.83 W m^{-2} °C^{-1}. The pizzas come out of the oven and go into the box at a temperature of 95 °C, and the outside air temperature is 12 °C. Calculate the thermal power loss from the carrying system. [3]

4.3 The pizza box has a temperature sensor inside. The sensor measures the temperature inside the pizza box every 10 minutes for 1 hour. The table shows this data.

Time/minutes	0	10	20	30	40	50	60
Temperature/°C	95	61	46	37	31	30	30

4.3.1 Plot the values in the table on a graph. Draw a line of best fit. [4]

4.3.2 On the same graph, sketch the line that you would expect if the pizza box was removed from the insulated carrying system. Explain your answer. [2]

Exam tip

When drawing graphs, ensure that:
+ the axes have a linear scale and an identified origin
+ the axes are labelled and have the correct units
+ the points are plotted correctly (within ±0.5 grid squares)
+ the best fit line is drawn as one smooth line, through the majority of the points.

---

5 Biofuels are an important future energy source. One form of biofuel involves turning waste vegetable oil from the catering industry (e.g. fish and chip restaurants) into biodiesel that can be used in diesel engines for vehicles.

5.1 Suggest why biodiesel is considered a **renewable fuel**. [2]

5.2 State **one** advantage and **one** disadvantage of using biodiesel to fuel vehicles. [2]

5.3 In a biodiesel engine, the useful power output as kinetic energy is 195 kW. The total power supplied by the biodiesel is 0.748 MW. Calculate the efficiency of the engine. [2]

5.4 The UK Government are phasing out combustion-based vehicles by 2035, so all new vehicles will have to be electric. This will help the UK to meet its Climate Change targets. Explain why biodiesel engines will not be allowed in the future. [2]

6 An electrical engineer is designing a de-mister for the back window of a car. They have designed some thin transparent resistor heating elements that can be glued to the inside of the window. One resistor, A, needs to be fitted to the small side window, and the others, B, C, D and E, will be fitted to the back window. The resistors can be arranged in two configurations, X and Y. Configuration X has resistor A in series with resistors B, C, D and E, which are connected in parallel with each other. Configuration Y has all five resistors connected in series. The battery has a voltage of 12.5 V.

Each resistor has a resistance of 8.0 Ω.

6.1 Calculate the total resistance of circuit Y. [1]

6.2 Calculate the current flowing through resistor A in circuit Y. [2]

6.3 Calculate the total resistance of circuit X. [2]

6.4 Calculate the current flowing through resistor A in circuit X. [2]

6.5 Compare the currents flowing through resistor B in both circuits. [4]

6.6 State which circuit, X or Y, will have the bigger rate of heat loss on the main window. [2]

7 The voltage–current values of two components, P and Q, are shown in the table below:

Current, I (A)		0.0	0.02	0.04	0.06	0.08	0.10
Voltage, V (V)	Component P	0.0	2.4	4.8	7.2	9.6	12.0
	Component Q	0.0	0.6	1.4	3.3	6.7	14.0

7.1 Draw a circuit diagram showing how you would obtain the data for component P. [4]

7.2 Plot a graph of the data and draw suitable lines of best fit. [4]

7.3 Identify component P and component Q. [2]

7.4 Calculate the resistance of component P. [2]

7.5 Use the graph to estimate the current and voltage values of component Q, when it has the same resistance as component P. [2]

7.6 Compare how the rate of heat loss from each component varies with the current. [2]

8 The circuit diagram below shows part of the control circuit for the solar panels on a satellite. Each solar panel (SP1 to SP5) is controlled by a controller with a resistance of 32 Ω.

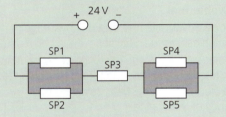

8.1 Calculate the total combined resistance of SP1 and SP2 in parallel. [1]

8.2 Calculate the overall resistance of the circuit. [1]

8.3 Calculate the current drawn from the power supply. [2]

8.4 SP3 is a heavier solar panel and requires a more powerful controller. Calculate the power of the SP3 controller. [2]

8.5 On board satellites, it is very important that all the conducting components are insulated from each other to prevent short circuits. Suggest a material that could be used to insulate the conductors. [1]

8.6 The solar panels are in use for 12 hours per day. Calculate the total charge that flows through controller SP3 in 12 hours. [2]

8.7 The controllers contain a mixture of metal conductor components and semiconductor components. Both types of component need to be insulated to prevent them from heating up too much or cooling down too much. Describe the effect of temperature on the resistance of each type of component. [2]

9 A student is designing a simple digital thermometer using a 14 Ω fixed resistor connected in series with a thermistor and a power supply. They also have a digital voltmeter and a beaker of hot water and a thermometer.

9.1 State and explain where the student should connect the voltmeter into the circuit so that when the temperature of the thermistor goes up, the voltage reading goes up. [2]

9.2 The figure shows how the resistance of the thermistor varies with temperature.

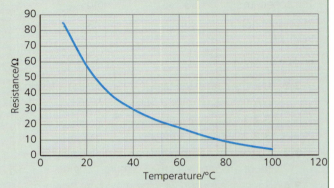

Copy the table and use the figure to complete the missing values. [1]

Temperature /°C	10	20	40	60	80	100
Current /A	0.10	0.09	0.07	0.05	0.03	0.01
Resistance of thermistor/Ω	85	58	30	18	9	4
Voltage across thermistor/V	8.5					0.04

9.3 Plot a VI graph of the data in the table. [4]

9.4 Use the data in the question to determine the voltage across the 14 Ω fixed resistor when the temperature is 20 °C. [3]

9.5 Calculate the voltage of the power supply. [2]

10 A mobile phone battery has a voltage of 3.79 V. It provides a power of 6.0 W to the screen and the processors connected in series.

10.1 Calculate the current drawn from the battery when the phone is in use. State the correct unit for current. [2]

10.2 The battery has a charge capacity of 4.29×10^4 C before the battery needs to be recharged. Calculate the time taken before the battery needs to be recharged if the phone is constantly drawing the current calculated in 10.1. [2]

10.3 The total resistance of the processors is 0.8 Ω. Calculate the voltage across the processors. [2]

10.4 Calculate the rate of heat loss from the processors. [2]

10.5 Suggest a reason why the charging time for a mobile phone battery is less when the charger has a higher voltage. [1]

11 A sports scientist is analysing the motion of a 0.450 kg rugby ball that is kicked off a kicking tee on the ground. The ball rises vertically, before reaching its highest point and falling back to the ground. She records the motion using a slow-motion video analysis system. She records the motion of the ball using a camera and then feeds the images into the analysing software.

11.1 She measures the initial velocity off the ground. Name the store of energy that the ball has immediately after it has been kicked. [1]

11.2 Explain, in terms of forces, why the ball remains stationary on the kicking tee before it is kicked. [2]

11.3 The ball rises to a height of 21 m in 2.2 seconds. Calculate the average speed of the ball during this time. [2]

11.4 The speed of the ball at the top of its flight is 0 m s⁻¹. Calculate the deceleration due to gravity and state its unit. [5]

11.5 Explain, in terms of energy, why the ball rises to a maximum height and then falls back to the ground. [3]

11.6 Calculate the gravitational potential energy of the ball at the top of its flight. [2]

12 During a car crash, air bags are deployed inside the car. These bags rapidly fill with air and cushion the driver and passengers from hard impacts with the dashboard or the front seats.

12.1 Explain, using the concept of momentum and Newton's Second Law, why air bags reduce injuries caused by car crashes. [4]

12.2 State the Law of Conservation of Momentum. [1]

12.3 During a low speed collision, a moving car of mass 1100 kg, travelling at 4.5 m s⁻¹, collides with a small stationary van of mass 1600 kg. The two vehicles lock together and move after the collision at a lower speed. Calculate the speed of the car/van combination immediately after the collision. [5]

13 During a sprint relay, the baton is handed over in a 20.0 m 'changeover' box. The recipient of the baton accelerates from rest so that she is sprinting at the same speed as the baton donor sprinter at the end of the box.

13.1 The donor sprinter is running at 9.0 m s⁻¹ inside the changeover box. Calculate the time taken for the donor sprinter to travel the length of the changeover box. [2]

13.2 The recipient sprinter needs to be running at the same speed as the donor sprinter at the end of the box, and she starts running as the donor sprinter just enters the box. Calculate the acceleration of the recipient sprinter. [3]

13.3 A velocity–time (v–t) graph of the donor sprinter's motion is shown in the figure. Calculate the acceleration of the donor sprinter. [2]

13.4 Use your value for acceleration to calculate the total distance travelled by the donor sprinter before the handover. [3]

13.5 The recipient sprinter sprints at 10 m s⁻¹ for most of the race but she decelerates at −2.3 m s⁻² for the last 2.5 seconds of the race. Calculate the distance travelled by the recipient sprinter during the last 2.5 s of the race. [2]

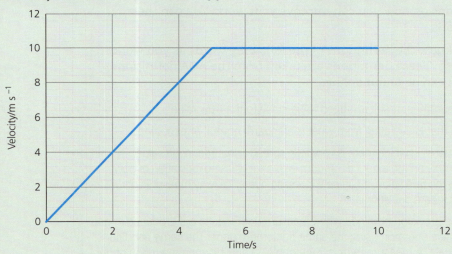

Introduction

This unit is assessed via internal assessment of a portfolio of assignments given to you by your centre. The assessment criteria are written so that you get credit for your experimental abilities and also your ability to research information. You need to practise your experimental techniques, taking guidance from your teachers or tutors.

It is not the purpose of this unit to produce a completed set of portfolio assignments for you to copy, but it will give you a checklist of things that you need to do, and give suggestions on how you might go about doing them. You must produce your own work, carried out independently. If you have worked as part of a group, this must be indicated at submission.

Your centre should provide you with an assignment brief, a copy of the grading criteria, a standard procedure for the experiment and a sample risk assessment (if appropriate – you will also have to produce one risk assessment in biology, one in chemistry and one in physics by yourself).

Your centre will also provide you with suitable deadlines to complete each assignment which must be followed. Once you have handed in your assignment, your teacher/tutor will give you some feedback focusing on what you have done well. Where you have not achieved a specific performance criterion or you want to improve a response, then your teacher/tutor can identify the issues related to the criterion. However, they will not provide explicit instructions on how you can improve your work. You can then resubmit your work.

You will need to know the definitions of the following words as you work through the course. These definitions are taken from the AQA definitions document, which can be found online.

Accuracy A measurement result is judged to be close to the true value.

Anomaly A value that is deviating from what is standard, normal or expected.

Precise Measurements in which there is very little spread about the mean value. Precision depends only on the extent of random errors – it gives no indication of how close results are to the true value. It can be expressed numerically by measures of imprecision (e.g. standard deviation).

Repeatability The precision obtained when measurement results are produced in one laboratory, by a single operator, using the same equipment under the same conditions, over a short timescale. A measurement is 'repeatable' in quality when repetition under the same conditions gives the same or similar results, e.g. when comparing results from the same learner or group using the same method and equipment.

Reproducibility The precision obtained when measurement results are produced by different laboratories (and therefore by different operators using different pieces of equipment). A measurement is 'reproducible' in quality when reproducing it under equivalent (but not identical) conditions gives the same or similar results from different learner groups, methods or equipment – a harder test of the quality of data.

Reliability The extent to which an experiment, test or measuring procedure yields consistent results on repeated trials. A measure is said to have a high reliability if it produces similar results under consistent conditions.

Validity The suitability of an investigative procedure to answer the question being asked.

Check your understanding and progress at **www.hoddereducation.co.uk/myrevisionnotes**

The portfolio assignments are as follows, and you need to produce one report for each technique:

Biology	Chemistry	Physics
1(a) Rate of respiration	2(a) Volumetric analysis	3(a) Resistivity
1(b) Light-dependent reaction in photosynthesis (the Hill reaction)	2(b) Colorimetric analysis	3(b) Specific heat capacity

Risk assessments

Performance outcome	Pass
	To achieve a pass the learner must evidence that they can:
PO4	**P10**
Understand safety procedure and risk assessment when undertaking scientific practical work	In using experimental techniques • safely use a range of practical equipment and materials • identify hazards • produce risk assessments for one applied experimental technique from each of biology, chemistry and physics.

If you are using an experiment to write a risk assessment:
+ You must produce a written risk assessment by yourself.
  + Do not use a numerical hazard/risk system.
  + Do not write your risk assessment in extended prose; use a table format.
+ You must use the format of risk assessment adopted by your centre. However, you should ensure that your risk assessments start (first column) with the identification of materials (chemicals, microorganisms, other materials, apparatus) and, where it is relevant, their state and concentration, name and type.
  + Apparatus can also be included, but 'glassware' and 'mains electrical equipment' can be one entry each. You do not need to detail the different types of glassware and main electrical equipment.
+ The next two columns should consider the hazards and risks of the identified materials. The nature of the hazard should correctly reflect the state and concentration of the material.
+ Further columns on the risk assessment should include:
  + control measures and PPE
  + disposal of materials if relevant
  + action on spillage/emergency or similar points.

> **Exam tip**
>
> The risk assessments that you produce should be in the same format as your other two written risk assessments and the risk assessments for the other three experiments provided by your centre.

Applied experimental techniques in biology

Rate of respiration

The AQA Performance outcomes for this section are shown below:

Performance outcome	Pass	Merit	Distinction
	To achieve a pass the learner must evidence that they can:	In addition to the pass criteria, to achieve a merit the learner must evidence that they can:	In addition to the pass and merit criteria, to achieve a distinction the learner must evidence that they can:
PO1(a)	**P2**	**M2**	**D2**

Performance outcome	Pass	Merit	Distinction
Rate of respiration	Follow a standard procedure to measure the effect of varying **one** given factor on the rate of respiration of a living organism.	Use formulae/calculations/ graphical representations to explain the data.	Evaluate the results and the method used.

Make sure you follow this checklist

You need to:

☐ describe the physiological measurements used in relation to rate of respiration in all organisms: the production of carbon dioxide and the uptake of oxygen

☐ indicate factors that affect the rate of respiration in all organisms:
   ☐ temperature
   ☐ concentration of glucose – this relates to experimental situations where simple organisms are fed with a glucose solution
   ☐ pH levels

☐ describe how the experiments treat organisms ethically

☐ detail the following methods for the physiological measurement and monitoring of cardiovascular, circulatory and respiratory systems in humans:
   ☐ peak flow
   ☐ lung capacity
   ☐ blood pressure (breathing rate and heart rate), including how this is used in the diagnosis of disease, improvement of performance in sport, or recovery from illness or injury

☐ list the factors that affect rate of respiration measurements in humans, as used by sport physiologists to determine metabolic rate while at rest and while exercising to ensure that energy expenditure meets energy inputs:
   ☐ metabolic rate, at rest and while exercising
   ☐ effects of given factors, such as temperature and exercise, on rate of respiration in a living organism

☐ give examples of the commercial and/or medical uses of physiological measurements.

☐ follow or design a risk assessment for this experiment
   ☐ Your centre will provide you with a risk assessment, or you should use their pro-forma for designing risk assessments.
   ☐ If you are designing a risk assessment it should be checked by a competent person (teacher/tutor) before you use it.
   ☐ If you are designing a risk assessment you should consult the relevant CLEAPSS Student Safety Sheets.
   ☐ You should check that you have included all the information needed for assessment based on the checklist on page 93.

☐ follow a given standard procedure (SP) to measure the effect of varying one factor on the rate of respiration in a living organism
   ☐ You may be given a standard procedure by your centre. An exemplar SP for this task is given below.
   ☐ Include photos of your experiment set-up and photos taken while the procedure is being carried out.
   ☐ You need to:
      ☐ describe the scientific principles of the technique, including (for the SP given here):
         ☐ the role of dehydrogenase enzyme in respiration
         ☐ the effect of temperature (or any other independent variable used) on enzymes
         ☐ the role of TTC as an alternative hydrogen acceptor (to those in respiration – FAD and NAD)
      ☐ safely and correctly set up and use the apparatus
      ☐ safely use the chemical reagent (e.g. TTC) involved in the experiment

Check your understanding and progress at **www.hoddereducation.co.uk/myrevisionnotes**

- obtain stopwatch readings recorded to the **nearest second** and record them correctly in a clear table, using the correct units (see below)
  - use formulae/calculations/graphical representations to explain the data.
- record all the relevant measurements
  - You need to consider:
    - precision
    - accuracy
    - reproducibility
    - relevant units for the measurements.
  - Make sure that you clearly measure (if appropriate) and record the following measurements (for the SP given below):
    - the mass of yeast
    - the concentration of the glucose solution
    - the volume of yeast suspension
    - the volume of TTC
    - the temperatures used – do not rely on the temperature setting of any thermostatically controlled water bath; instead, use a thermometer to read the temperature
    - the time taken for the colour change.

Exam tip

Temperature intervals do not have to be 'rounded' temperatures (20 °C, 30 °C, 40 °C, etc.) as long as the temperatures are accurately recorded and the intervals are roughly the same.

AQA specify the definitions of the terms: precise, accurate and reproducible. You should refer to these definitions on page 92.

Exemplar standard procedure

Investigating the effect of temperature on the rate of respiration in yeast

Scope: This SP can be used to measure the effect of temperature on the respiration rate of microorganisms. By changing the independent and control variables, it could also be used to measure the effect of pH or glucose concentration.

Principle: Dehydrogenase enzyme is important in respiration as it catalyses the removal of hydrogen for transfer to NAD or FAD. In this experiment, the TTC acts as an alternative hydrogen acceptor. When reduced, TTC changes from a colourless liquid to pink. The time for this colour change to occur can be used to measure the rate of respiration.

Materials needed: Dried active yeast; distilled water; glucose; 0.5% TTC solution

Equipment needed: Balance; spatula; 5 × 250 cm³ conical flasks; glass rod; 100 cm³ measuring cylinder; 5 × water baths at different temperatures; thermometer; stopwatch (accurate to ± 1 second); 10 cm³ graduated pipette; 1 cm³ syringe; 10 × test tubes, incubator

Safety: A risk assessment compatible with your centre should be written or followed. Written risk assessments must be checked by a competent person before use. TTC is an irritant. Wear eye protection.

Procedure:

1. Add 10 g of dried yeast to 100 cm³ of distilled water in a conical flask.
2. Add 2.5 g of glucose to the flask and stir thoroughly using a glass rod.
3. Place the flask in an incubator for 1 hour.
4. Using a 1 cm³ syringe, add 1 cm³ of TTC to each of five test tubes.
5. Add 10 cm³ of yeast suspension to each of the five remaining test tubes using a graduated pipette.
6. Place one test tube of yeast and one of TTC in each of the five water baths and leave for a minimum of 5 minutes to equilibrate.
7. After equilibration, add the TTC solution from one of the water baths to the yeast suspension from the same water bath and use a stopwatch to measure the time (in seconds) until the TTC turns pink.
8. Repeat step 7 for the tubes from the other water baths.

Calculations: The rate of respiration can be expressed using the following equation:

$$\text{Rate of respiration} = \frac{1000}{\text{time in seconds}}$$

The top of this fraction can be any constant. 1000 is likely to give manageable figures in this experiment.

Expression of results: Record the results in a table and then plot a line graph (with temperature on the x-axis and the calculated rate of respiration on the y-axis).

- Analyse the results to evaluate the methodology of your experiment.
  - Compare your results with those of other groups to assess reproducibility.
  - Consider whether your results show a clear trend and whether any such trend is consistent with theories of enzyme action.
- Include details of how the organisms used (yeast, in this case) were treated ethically. With single-celled microorganisms, ethical considerations will be limited to avoiding the death of the organisms.

You need to explain the importance of using measurements of respiration rate in an industrial/commercial context. Yeast is used in the brewing and baking industries and in the production of ethanol biofuel.

Light-dependent reaction in photosynthesis (the Hill reaction)

REVISED ●

The AQA Performance outcomes for this section are shown below:

Performance outcome	Pass	Merit
	To achieve a pass the learner must evidence that they can:	In addition to the pass criteria, to achieve a merit the learner must evidence that they can:
PO1(b)	**P3**	**M3**
The light-dependent reaction in photosynthesis (the Hill reaction)	Follow a standard procedure to measure the Hill reaction and record results.	Explain how this standard procedure could be adapted to investigate **three** limiting factors.

Make sure you follow this checklist

You need to:
- ☐ state the definition and equations for the overall process of photosynthesis
- ☐ describe and understand the light-dependent and light-independent reactions in photosynthesis
- ☐ make measurements, calculations and graphical presentations of photosynthesis
- ☐ describe the factors affecting photosynthesis:
  - ☐ light intensity
  - ☐ light wavelength
  - ☐ carbon dioxide
  - ☐ temperature
  - ☐ chlorophyll availability
  - ☐ herbicide inhibitors
  - ☐ plant species.
- ☐ detail the science related to the limiting factors in photosynthesis
- ☐ explain the value of manipulating factors to increase productivity
- ☐ list the criteria for suitable specimens
- ☐ follow a given standard procedure (SP) to measure the Hill reaction
  - ☐ You may be given a standard procedure by your centre. An exemplar SP for this task is given below.
  - ☐ Include photos of your experiment set-up and photos taken while the procedure is being carried out.
  - ☐ You need to:
    - ☐ describe the scientific principles of the technique
    - ☐ explain how the standard procedure for the Hill reaction could be adapted to investigate three of the factors listed above
    - ☐ safely and correctly set up and use the apparatus
    - ☐ safely use the chemical reagent (e.g. DCPIP) involved in the experiment
    - ☐ obtain readings recorded to the nearest second for the time taken for the DCPIP to decolourise and record them correctly in a clear table, using the correct units.
- ☐ record all the relevant measurements.
  - ☐ You need to consider:
    - ☐ precision
    - ☐ accuracy
    - ☐ reproducibility
    - ☐ relevant units for the measurements.

Check your understanding and progress at **www.hoddereducation.co.uk/myrevisionnotes**

- Make sure that you clearly measure (if appropriate) and record the following measurements (for the SP given below):
  - the concentration of the DCPIP
  - the volumes of isolation medium used
  - the volume of DCPIP
  - the time taken for the colour change.

> AQA specify the definitions of the terms: precise, accurate and reproducible. You should refer to these definitions on page 92.

Exemplar standard procedure

Measuring the Hill reaction using DCPIP

This SP is based on the SP produced by the Royal Society of Biology and the Nuffield Foundation.

Scope: This SP can be used to measure the rate of the light-dependent reaction in any green plant.

Principle: DCPIP (2,6-dichlorophenol-indophenol), a blue dye, acts as an electron acceptor and becomes colourless when reduced, allowing electrons produced in the light-dependent stage of photosynthesis to be detected.

Materials needed: Fresh green spinach, lettuce or cabbage, 3 leaves (discard the midribs); ice-cold 0.05 M phosphate buffer solution, pH 7.0; ice-cold isolation medium (sucrose and KCl in phosphate buffer); potassium chloride; DCPIP solution (1×10^{-4} mol dm^{-3} approx.)

Equipment needed: Centrifuge – with RCF between 1500 and 1800g; centrifuge tubes; scissors; cold pestle and mortar (or blender or food mixer) which has been kept in a freezer compartment for 15–30 minutes (if left too long the extract will freeze); muslin or fine nylon mesh; filter funnel; ice-water salt bath; glass rod or Pasteur pipette; measuring cylinder, 20 cm^3; beaker, 100 cm^3; pipettes, 5 cm^3 and 1 cm^3; bench lamp with 100 W bulb; test tubes × 5; boiling tube; pipette filler; waterproof pen to label tubes; colorimeter and tubes or light sensor and data logger

Safety: A risk assessment compatible with your centre should be written or followed. Written risk assessments must be checked by a competent person before use. DCPIP is harmful. All other chemicals used are low hazard. Wear eye protection.

Procedure: Keep solutions and apparatus cold during the extraction procedure, steps 1–8, to preserve enzyme activity. Carry out the extraction as quickly as possible.

Preparation:

1. Cut three small green spinach, lettuce or cabbage leaves into small pieces with scissors, discarding the midribs and leaf stalks. Place in a cold mortar or blender containing 20 cm^3 of cold isolation medium. (Larger quantities will be needed if a blender is used.)

2. Grind vigorously and rapidly (or blend for about 10 seconds).

3. Place four layers of muslin or nylon in a funnel and wet with cold isolation medium.

4. Filter the mixture through the funnel into the beaker and pour the filtrate into pre-cooled centrifuge tubes supported in an ice-water salt bath. Gather the edges of the muslin, wring thoroughly into the beaker and add filtrate to the centrifuge tubes.

5. Check that each centrifuge tube contains about the same volume of filtrate.

6. Centrifuge the tubes for sufficient time to get a small pellet of chloroplasts (10 minutes at high speed should be sufficient).

7. Pour off the liquid (supernatant) into a boiling tube, being careful not to lose the pellet. Re-suspend the pellet with about 2 cm^3 of isolation medium, by stirring with a glass rod. Squirting in and out of a Pasteur pipette five or six times will give a uniform suspension.

8. Store this leaf extract in an ice-water salt bath and use as soon as possible.

Investigation using the chloroplasts

Read all the instructions before you start. Use the DCPIP solution at room temperature.

Set up five labelled tubes as follows.

Test tube	Leaf extract (cm^3)	Supernatant (cm^3)	Isolation medium (cm^3)	Distilled water (cm^3)	DCPIP solution (cm^3)
1	0.5	–	–	–	5
2	–	–	0.5	–	5
3	0.5	–	–	–	5
4	0.5	–	–	5	
5	–	0.5	–	–	5

1 When the DCPIP is added to the extract, shake the tube and note the time. Place tubes 1, 2 and 4 about 12–15 cm from a bright light (100 W). Place tube 3 in darkness.

2 Time how long it takes to decolourise the DCPIP in each tube. If the extract is so active that it decolourises within seconds of mixing, dilute it 1:5 with isolation medium and try again.

Expression of results: Record the results in a table.

You need to explain how you would modify the SP to investigate the effect of three factors on the rate of the Hill reaction (i.e. the light-dependent stage). You should include the modifications to the method, the apparatus used and its arrangement, readings to be taken, and some scientific background.

Exam tip

In explaining the investigation of three factors and their effect on the light-dependent stage, you must use a modification of the SP already carried out, not some other technique. The scientific background to the modifications should be explained.

Applied experimental techniques in chemistry

Volumetric analysis

REVISED ●

The AQA Performance outcomes for this section are shown below:

Performance outcomes	Pass	Merit	Distinction
	To achieve a pass the learner must evidence that they can:	In addition to the pass criteria, to achieve a merit the learner must evidence that they can:	In addition to fulfilling the pass and merit criteria, to achieve a distinction the learner must evidence that they can:
PO2 Demonstrate applied experimental techniques in chemistry	**P4** Outline the basic principles and uses of volumetric analysis. (This is assessed in conjunction with 2(b) Colorimetry.)	**M4** Explain the scientific principles of volumetric analysis with reference to: + standard solutions + choice of indicators. (This is assessed in conjunction with 2(b) Colorimetry.)	
PO2(a) Volumetric analysis	**P5** Follow a standard procedure for volumetric analysis by: + preparing a standard solution + carrying out a titration + recording all measurements and data.	**M5** Carry out calculations that support: + preparation of the standard solution + the titration.	**D3** Explore how the technique is used in industry, with reference to accuracy and the use of primary standards.
If applicable: **PO4** Understand safety procedure and risk assessment when undertaking scientific practical work. (This is assessed if the Risk assessment has been given to you in 2(b) Colorimetry.)	**P10** In using experimental techniques: + safely use a range of practical equipment and materials + identify hazards + produce risk assessments for one applied experimental technique from each of biology, chemistry and physics.		

Check your understanding and progress at **www.hoddereducation.co.uk/myrevisionnotes**

Make sure you follow this checklist

In your portfolio you need to include the following:

- [ ] Describe the volumetric analysis (titration) techniques used in industry. You should consider their convenience and accuracy in determining the amount of substance present in a sample. You need to cover:
    - [ ] the uses, applications and scientific principles of volumetric analysis, considering:
        - [ ] the reaction, equation, stoichiometry and equivalence/mole ratios
        - [ ] the choice of indicator (if required) and the colour change at the end point
            - [ ] for acid–base reactions, you need to justify the choice of indicator based on the reaction and pH curve
            - [ ] for other types of reaction, you should describe how the colour change at the end point is achieved/occurs
        - [ ] the importance of using a primary standard to prepare standard solutions or a sample that has been standardised. You also need to detail their use in determining unknown concentrations. You should describe the properties of a primary standard.
- [ ] Detail the types of titration where an acid and a soluble base or alkali are titrated together in the presence of a suitable indicator:
    - [ ] determination of acid in rainwater
    - [ ] ethanoic acid in vinegar
    - [ ] lactic acid in milk
    - [ ] acid in de-scalers
    - [ ] bases (e.g. used in drain or oven cleaners).
- [ ] Detail titrations involving other types of reaction such as:
    - [ ] precipitation, such as the determination of the concentration of chloride ions in foodstuffs or polluted water
    - [ ] reduction/oxidation (redox), such as the concentration of vitamin C in foodstuffs
    - [ ] the formation of metal complex ions, such as the concentration of metal ions in polluted river water.
- [ ] Describe how volumetric analysis is used in analytical laboratories.
- [ ] Follow or design a risk assessment for this experiment.
    - [ ] Your centre will provide you with a risk assessment, or you should use their pro-forma for designing risk assessments. You should have evidence of the risk assessment in your portfolio.
    - [ ] If you are designing a risk assessment it should be checked by a competent person (teacher/tutor) before you use it.
    - [ ] If you are designing a risk assessment you should consult the relevant CLEAPSS Student Safety Sheets.
    - [ ] You should check that you have included all the information needed for assessment based on the checklist on page 93.
- [ ] Follow a given standard procedure (SP) to determine the concentration of a sample of river water believed to be contaminated with hydrochloric acid.
    - [ ] You may be given a standard procedure by your centre. An exemplar SP for this task is given below. You should have evidence of the standard procedure in your portfolio.
    - [ ] Include photos of your experiment set-up and photos taken while the procedure is being carried out.
    - [ ] You need to:
        - [ ] describe the scientific principles of the technique, including:
            - [ ] the correct formula for the reaction
            - [ ] the stoichiometry of the reaction
        - [ ] use a primary standard to prepare a standard solution of sodium hydroxide
            - [ ] calculate the number of moles of any solids used

Making links

In Unit 1 you met the concepts of: moles; molarity; stoichiometry and formulae; pH curves; end points; the choice of indicator; and titration calculations based on volumes and concentration.

Exam tip

You **do not** need to know about other forms of titration such as *thermometric* and *potentiometric*.

Exam tip

Your risk assessment **must** include the states (phases) and concentrations of any chemicals used, and an assessment of the risks taking these factors into account.

- make a standard solution of sodium hydroxide using your primary standard
- correctly determine the concentration (M) of your standard solution
- safely and correctly set up and use the apparatus
- safely use the chemical reagents involved in the preparation of the standard solution and the titration
- obtain burette readings recorded to the *nearest 0.05 cm³* and record them correctly in a clear table, using the correct units (see below)
- find and recognise the end point of the reaction with a suitable indicator
- repeat your measurements until you get reproducible, concordant results (titres within ±0.10 cm³)
- calculate a mean titre.
- You must record all the relevant measurements.
  - You need to consider:
    - precision
    - accuracy
    - reproducibility
    - concordant titres
    - relevant units for the measurements.
  - Make sure that you clearly measure and record the following measurements:
    - the mass of the primary standard measured out
    - the mass of the solid sodium hydroxide used to produce the standard solution
    - the volume of your samples
    - the titration data, including the initial and final burette readings and titres.

> **Exam tip**
>
> Your titration should be carried out individually, and the results should be yours.

> **Exam tip**
>
> In order to meet performance outcome P5, you must design *your own clear table* of data. You cannot use a template or a pro-forma.

> AQA specify the definitions of the terms: precise; accurate and reproducible. You should refer to these definitions on page 92.

Exemplar standard procedure

Preparing a standard solution of sodium hydroxide using a primary standard

This SP is adapted from the SP produced by Practical Chemistry: Making a standard solution and the Royal Society of Chemistry: Resources: Standard solutions.

Scope: This SP can be used to produce a dilute (~0.1 mol dm⁻³) standard solution of sodium hydroxide for use in a titration.

Principle: Potassium hydrogen phthalate, KHP, is a primary standard and is often used to standardise solutions of sodium hydroxide. A primary standard solution of KHP is produced first, followed by a solution of sodium hydroxide. The KHP solution is then used to accurately determine the concentration of the sodium hydroxide by volumetric analysis.

Materials needed: Solid potassium hydrogen phthalate (KHP); solid sodium hydroxide pellets; distilled water; phenolphthalein indicator

Equipment needed: Electronic laboratory balance (±0.01 g or less); 2 × weighing boats; 2 × spatulas; 2 × 250 cm³ glass beakers; wash bottle containing distilled water; 2 × glass stirring rods; 2 × 250 cm³ volumetric flasks; 2 × glass funnels to fit volumetric flasks; 2 × dropping pipettes

Safety: A risk assessment compatible with your centre should be written or followed. Written risk assessments must be checked by a competent person before use. You should consult the relevant CLEAPSS Student Safety Sheets: 31 Sodium hydroxide; 70 Dyes and indicators. Potassium hydrogen phthalate is an irritant. Wear eye protection.

Part A – Preparation of the KHP primary standard solution
Procedure:
1 Place a weighing boat onto the electronic balance and tare the balance.
2 Transfer 5.00 g of KHP onto the weighing boat. Measure and record the mass of KHP.
3 Pour the KHP into a 250 cm³ glass beaker.
4 Use distilled water from the wash bottle to ensure that all the KHP is washed out of the weighing boat into the beaker.
5 Add 50 cm³ of distilled water to the beaker and stir the solution to completely dissolve the KHP, adding more distilled water as required
6 Transfer the solution to the volumetric flask through the glass funnel.
7 Rinse the beaker and stirring rod with more distilled water into the volumetric flask, ensuring that all of the solution has been rinsed into the volumetric flask.
8 Fill up the volumetric flask to within 1 cm of the graduation line. Remove the funnel.

9 Use the dropping pipette to add more distilled water to the flask so that the bottom of the meniscus is level with the graduation line. Add the stopper and invert the volumetric flask ten times to ensure that the solution is completely mixed.

10 Label the flask, including the mass of KHP added. Leave space to write the concentration of the KHP solution.

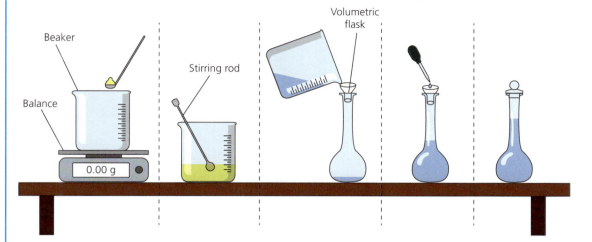

Figure 2.1 Equipment for preparing a standard solution

Calculations: The molar mass of potassium hydrogen phthalate is $204.22\,g\,mol^{-1}$. Use this value, the mass of KHP added to the solution and the volume of the volumetric flask to calculate the concentration of the KHP solution.

Expression of results: Write the concentration of the KHP solution on the side of the volumetric flask.

Part B – Preparation of the sodium hydroxide standard solution

Procedure:

1 Place a weighing boat onto the electronic balance and tare the balance.

2 Transfer 1.00 g of sodium hydroxide pellets onto the weighing boat. Record the mass of sodium hydroxide.

3 Pour the sodium hydroxide into a $250\,cm^3$ glass beaker.

4 Repeat steps 4–10 from Part A.

5 Use a volumetric pipette to add $25.0\,cm^3$ of your standard sodium hydroxide solution to a conical flask. Add two drops of phenolphthalein indicator and titrate with your primary standard solution of KHP using the standard titration procedure outlined below.

Calculations: Use the concentration of the KHP, the volume of the standard solution of sodium hydroxide and the mean concordant titres to calculate the concentration of the standard solution of sodium hydroxide.

Expression of results: Write the concentration of the sodium hydroxide standard solution, using the correct units, on the side of the volumetric flask.

Exemplar standard procedure

Performing a titration to determine the concentration of river water polluted with hydrochloric acid

This SP is adapted from the SP produced by Wired Chemist: wiredchemist.com/chemistry/instructional/laboratory-tutorials/volumetric-analysis and *AQA A-level Chemistry*; McFarland, A., Quigg, T. and Henry, N.; Hodder Education 2019.

Scope: This SP can be used to determine the concentration of a river water sample contaminated with hydrochloric acid with a concentration of approximately $0.1\,mol\,dm^{-3}$, using a sodium hydroxide standard solution.

Principle: The acidic water sample is neutralised by a standard solution of sodium hydroxide. A suitable indicator is used to determine the end point of the titration.

Materials needed: Sodium hydroxide standard solution; river water sample; indicator suitable for the reaction.

Equipment needed: $50\,cm^3$ burette; burette clamp and stand; white tile; $250\,cm^3$ conical flask; $25\,cm^3$ volumetric pipette and filler; glass funnel to fit the burette; black tape strip; $250\,cm^3$ beaker.

Safety: A risk assessment compatible with your centre should be written or followed. Written risk assessments must be checked by a competent person before use. You should consult the relevant CLEAPSS Student Safety Sheets: 20 Hydrochloric acid; 31 Sodium hydroxide; 70 Dyes and indicators. Wear eye protection.

Procedure:

1 A diagram summarising how to set up and perform a titration is shown in Figure 2.2.

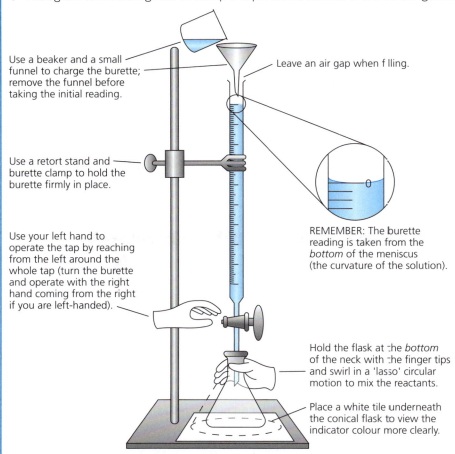

Use a beaker and a small funnel to charge the burette; remove the funnel before taking the initial reading.

Leave an air gap when filling.

Use a retort stand and burette clamp to hold the burette firmly in place.

REMEMBER: The burette reading is taken from the *bottom* of the meniscus (the curvature of the solution).

Use your left hand to operate the tap by reaching from the left around the whole tap (turn the burette and operate with the right hand coming from the right if you are left-handed).

Hold the flask at the *bottom* of the neck with the finger tips and swirl in a 'lasso' circular motion to mix the reactants.

Place a white tile underneath the conical flask to view the indicator colour more clearly.

Figure 2.2 Carrying out a titration

2 Rinse the burette with distilled water. Then rinse with approximately 5 cm³ of the standard sodium hydroxide solution **three** times. The standard sodium hydroxide solution is called the titrant.

3 Fill the burette with titrant using the glass funnel, until the solution has gone above the 0.0 cm³ line.

4 Remove the funnel. Place a beaker under the end of the burette and adjust the volume of titrant to the 0.0 cm³ line. Avoid parallax error by measuring the line to the bottom of the titrant meniscus at eye level, and ensuring that the jet below the tap is filled with solution, with no air bubble. Use the piece of black tape (or hold a white tile behind the meniscus) as a guideline. Discard the titrant in the beaker.

5 Use a clean, dry volumetric pipette to transfer 25.0 cm³ of the acidic river water sample to a clean, dry 250 cm³ conical flask.

6 Identify the appropriate indicator to use with the titrant and the sample. Add three to five drops of the appropriate indicator to the sample and place the conical flask onto a white tile.

7 Add the titrant to the sample, swirling with a lasso motion as shown in Figure 2.2. The indicator will change colour at the end point of the reaction.

8 The first titration should be rough and should be an overshoot, so this titre will be greater in value than the three subsequent accurate titrations.

9 The subsequent titrations should be concordant (within 0.10 cm³ of each other) with drop-wise addition as the end point is reached. Titrations which are not concordant (within 0.10 cm³ of each other) are recorded but not used to calculate the average titre.

10 When the end point has been reached, allow the titrant to sit in the burette for 10 seconds to allow the liquid to settle, then measure the volume to the nearest 0.05 cm³, avoiding parallax errors as in step 4.

11 Repeat the titration until you have three concordant titres.

12 A standard burette contains 50.0 cm³ of titrant. If the rough titre of titrant is close to or above 25 cm³, you should refill the burette between titrations.

13 Record all titration volume measurements in a suitable table.

Calculations: Use the concentration of the standard sodium hydroxide solution, the volume of the acidic river water sample and the mean concordant titres to calculate the concentration of the acidic river water.

Expression of results: Record the concentration of your acidic river water sample using the correct units.

Check your understanding and progress at **www.hoddereducation.co.uk/myrevisionnotes**

- ☐ Calculate a value for the concentration of the river water.
  - ☐ Perform suitable calculations using your recorded results and the formula linking concentration and volume from page 43. You will need to do this twice:
    - ☐ once for the standard solution of sodium hydroxide, using the mass of the primary standard
    - ☐ once for the unknown polluted river water.
  - ☐ Suggest a percentage error in your calculated value of the concentration of the polluted river water.
- ☐ Fully evaluate the methodology of your experiment.
  - ☐ Compare your calculated concentration value with a value determined by your teacher/tutor/technician.
    - ☐ Compare the levels of precision, accuracy and reliability of these two values.
    - ☐ Suggest reasons for any discrepancies between these two values.
  - ☐ Research the methods used in industry and analytical laboratories. Explain the use of:
    - ☐ auto-pipettes
    - ☐ auto-titrators
    - ☐ electronic sensors or electrodes.
  - ☐ Compare the precision, accuracy and reproducibility of results obtained by industrial/analytical laboratory techniques with those obtained by you using standard school/college laboratory glassware.
  - ☐ Explain the importance of using primary standards in industrial/analytical laboratories to ensure the accuracy of measurements and calculated values of concentration.

> **Exam tip**
>
> To meet performance outcome M5, you must perform all calculations without using a template or scaffolding.

> **Exam tip**
>
> You should include diagrams and images of the apparatus used in industry/analytical laboratories.

> **Exam tip**
>
> Make sure that you fully reference any researched information that you use in your report.

Colorimetric analysis

REVISED ⭕

The AQA Performance outcomes for this section are shown below:

Performance outcomes	Pass	Merit	Distinction
	To achieve a pass the learner must evidence that they can:	In addition to the pass criteria, to achieve a merit the learner must evidence that they can:	In addition to fulfilling the pass and merit criteria, to achieve a distinction the learner must evidence that they can:
PO2 Demonstrate applied experimental techniques in chemistry	**P4** Outline the basic principles and uses of colorimetry. (This is assessed in conjunction with 2(a) Volumetric analysis.)	**M4** Explain the scientific principles of colorimetry with reference to: consideration of the Beer–Lambert Law. (This is assessed in conjunction with 2(a) Volumetric analysis.)	
PO2(b) Colorimetric analysis	**P6** Follow a standard procedure for colorimetric analysis, using solution dilutions, by: + recording all data and measurements + producing a calibration curve + determining the unknown concentration.	**M6** Explain the choice of filter/wavelength, describe any inconsistencies in the data recorded making reference to the Beer–Lambert Law.	**D4** Evaluate the outcome of analysis with reference to precision, reliability and accuracy.

103

Performance outcomes	Pass	Merit	Distinction
If applicable: **PO4** Understand safety procedure and risk assessment when undertaking scientific practical work. (This is assessed if the Risk assessment has been given to you in 2(a) Volumetric analysis.)	**P10** In using experimental techniques: + safely use a range of practical equipment and materials + identify hazards + produce risk assessments for one applied experimental technique from each of biology, chemistry and physics.		

Make sure you follow this checklist

You need to be able to do the following:

☐ Describe the scientific principles of colorimetry. This should include written descriptions of:
  - ☐ how colorimetry depends on the absorption of a particular wavelength of light by a coloured solution
  - ☐ the visible electromagnetic spectrum, electron energy levels and the mechanism of absorption, including the nature of the colour of some compounds
  - ☐ the construction of a simple colorimeter, which must include: the light source; the use of coloured filters or a means of selecting wavelength; the cuvette and the path length of the light; the detection of the transmitted light and the display of the measured value (you should include a diagram)
  - ☐ the choice of filter or wavelength of incident light and the use of absorption against wavelength curves and the wavelength of maximum absorption, λ_{max}
  - ☐ the calibration of the colorimeter by standard solutions and the nature of the absorbance against concentration calibration graph (you should include an example calibration graph)
  - ☐ the Beer–Lambert Law, which links the absorbance of a solution to the path length of light through the solution and the concentration of the solution, and the application of the Beer–Lambert Law to the calibration graph of absorbance against concentration.

☐ Describe the use of colorimetric techniques in industry and analytical laboratories, to measure the concentrations of coloured solutions. You need written descriptions of:
  - ☐ the analysis of metal ion concentrations in ecological samples, e.g. for pollution analysis
  - ☐ the quality control of products containing metals such as copper and iron
  - ☐ how colorimetric techniques are used in the dye/pigment/ink industry and the food/beverage industry.

☐ Describe the use of colorimetry to analyse colourless substances using suitable complexing agents, which produce a coloured solution when mixed with a colourless sample for analysis.

☐ Follow or design a risk assessment for this experiment. You must have evidence of this in your portfolio.
  - ☐ Your centre will provide you with a risk assessment, or you should use their pro-forma for designing risk assessments.
  - ☐ If you are designing a risk assessment it should be checked by a competent person (teacher/tutor) before you use it.
  - ☐ If you are designing a risk assessment you should consult the relevant CLEAPSS Student Safety Sheets.
  - ☐ You should check that you have included all the information needed for assessment based on the checklist on page 93.

> **Exam tip**
>
> Colorimeters will also measure the *transmission* of light through a solution. AQA are expecting that you will use *absorbance* throughout this work, both in terms of the underlying principles and in practical work.

> **Making links**
>
> You will find this experiment easier if you go back through the work that you did in Unit 1 on atomic structure; the amount of substance; and electron transitions brought about by the absorption of light of a specific wavelength.

> **Exam tip**
>
> Your risk assessment **must** include the states (phases) and concentrations of any chemicals used, and an assessment of the risks taking these factors into account.

Check your understanding and progress at **www.hoddereducation.co.uk/myrevisionnotes**

- Follow a given standard procedure (SP) to determine the concentration of a sample of a known coloured solution, such as copper(II) sulfate. You must have evidence of this in your portfolio.
  - You may be given a standard procedure by your centre. An exemplar SP for this task is given below.
  - Include photos of your experimental set-up and photos taken while the procedure is being carried out.
  - You need to:
    - use a standard procedure to determine the concentration of a sample of coloured solution
    - identify the wavelength of light for maximum absorption (λ_{max}) or the most appropriate filter to use; record the absorbance values for each filter (or wavelength)
    - prepare and use a standard solution to calibrate the colorimeter
    - state the equipment used
    - plot an absorption curve for the substance investigated.
- You must record all the relevant measurements.
  - You need to consider the appropriate level of precision and the correct relevant units to use for all measurements.
  - You should include the measurements taken preparing the standard solutions and the calculations needed to determine required concentrations.
  - You should include your measurements for the selection of the chosen filter as well as the absorbances for each known concentration and the unknown sample.
  - You should measure and record the value of absorbance using distilled water prior to zeroing as evidence that you have included this step.
- You need to process and interpret your results to determine the concentration of the unknown test sample. This should include:
  - a graph of absorbance against wavelength (if appropriate) or a chart of absorbance against filter colour
  - a calibration graph of absorbance against concentration, used to determine the concentration of the unknown sample. You should:
    - assess any inconsistencies/anomalies in your absorbance data (you could plot error bars using the range of each measurement – if appropriate)
    - assess how well the line of best fit complies with the Beer–Lambert Law.
  - a comparison of your value of the unknown concentration with the value determined by your teacher/tutor/technician.

Exam tip

Your colorimetric analysis should be carried out individually, and the results should be yours.

Exam tip

To meet performance outcome P5, you must design *your own clear table* of data. You cannot use a template or a pro-forma.

Exam tip

Ensure that all graphs/charts have the correct scale, correctly plotted points and suitable lines of best fit (if appropriate). You can use Excel to plot your graphs but you should ensure that the graphs have all the required elements as Excel, by default, will produce only a basic graph to start with, which needs editing to get it into the correct AQA format. If in doubt, plot any graphs or charts by hand on suitable graph paper.

- You need to evaluate the methodology used in this experiment. You need to consider and include in your portfolio:
  - any quantitative (%) errors associated with all your measurements
  - the precision of recording of your measurements
  - the reliability of your measurements
  - a qualitative assessment of your practical methodology
  - an assessment of the accuracy of your final value for the unknown concentration, considering your error analysis.

AQA specify the definitions of the terms precise, accurate and reproducible. You should refer to these definitions on page 92.

Exemplar standard procedure

Preparing a standard solution of copper(II) sulfate

This SP is adapted from the SP produced by Practical Chemistry: Making a standard solution and the Royal Society of Chemistry: Resources: Standard solutions.

Scope: This SP can be used to produce a 1.0 mol dm⁻³ standard solution of copper(II) sulfate for use in the colorimetric analysis of an unknown concentration of copper(II) sulfate.

Principle: Copper(II) sulfate is available as crystals of copper(II) sulfate pentahydrate, $CuSO_4.5H_2O$. A standard 1.0 mol dm⁻³ solution is prepared, followed by serial dilutions to produce a range of solutions between 0.1 mol dm⁻³ and 1.0 mol dm⁻³. These are then used in a separate SP to produce a calibration curve to determine the unknown concentration of a sample of copper(II) sulfate solution.

Materials needed: Solid copper(II) sulfate pentahydrate; distilled water

Equipment needed: Electronic laboratory balance (±0.01 g or less); weighing boats; spatula; 250 cm³ glass beaker; wash bottle containing distilled water; glass stirring rod; 250 cm³ volumetric flask; glass funnel to fit volumetric flask; dropping pipette; 2 × 10 cm³ graduated pipettes and fillers; 10 × test tubes; test tube racks to fit 10 test tubes.

Safety: A risk assessment compatible with the centre should be written or followed. Written risk assessments must be checked by a competent person before use. You should consult the relevant CLEAPSS Student Safety Sheets: 40 Copper and its compounds. Copper(II) sulfate is harmful. Wear eye protection.

Part A – Preparation of the copper(II) sulfate standard solution

Procedure:

1 Use the formula of copper(II) sulfate pentahydrate to calculate its relative formula mass, RFM. You will be making up a 1.0 mol dm⁻³ solution in a 250 cm³ volumetric flask. Use this information to determine the mass of copper(II) sulfate pentahydrate needed in the flask, X g.

2 Transfer X g of copper(II) sulfate pentahydrate onto the weighing boat. Measure and record the mass of copper(II) sulfate pentahydrate.

3 Pour the copper(II) sulfate pentahydrate into a 250 cm³ glass beaker.

4 Use distilled water from the wash bottle to ensure that all the copper(II) sulfate pentahydrate has been washed out of the weighing boat into the beaker.

5 Add 50 cm³ of distilled water to the beaker and stir the solution to completely dissolve the copper(II) sulfate pentahydrate, adding more distilled water as required.

6 Transfer the solution to the volumetric flask through the glass funnel.

7 Rinse the beaker and stirring rod with more distilled water into the volumetric flask, ensuring that all of the solution has been rinsed into the volumetric flask.

8 Fill up the volumetric flask to within 1 cm of the graduation line. Remove the funnel.

9 Use the dropping pipette to add more distilled water to the flask so that the bottom of the meniscus is level with the graduation line. Add the stopper and invert the volumetric flask ten times to ensure that the solution is completely mixed.

10 Label the flask, including the mass of copper(II) sulfate pentahydrate added. Leave space to write the concentration of the copper(II) sulfate solution.

Calculations: Determine the mass, X g, of copper(II) sulfate pentahydrate needed in the flask using Step 1 of the procedure.

Expression of results: Write the concentration of copper(II) sulfate solution on the side of the volumetric flask.

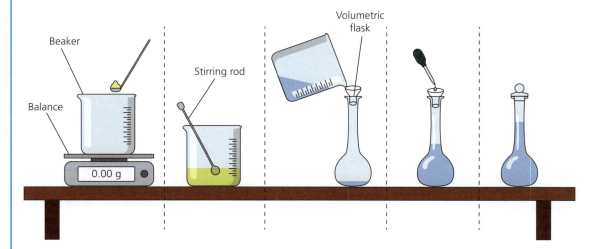

Figure 2.3 Equipment for preparing a standard solution

Part B – Serial dilution of standard copper(II) sulfate

Procedure:

1. Use a 10 cm³ graduated pipette to transfer 10.0 cm³ of your 1.0 mol dm⁻³ copper(II) sulfate solution into a test tube. Tap the pipette against the inside of the test tube to ensure that all of the solution has been transferred. Label the test tube as 1.0 mol dm⁻³.

2. Use the 10 cm³ graduated pipette to transfer 9.0 cm³ of your 1.0 mol dm⁻³ copper(II) sulfate solution into a second test tube. Use the second graduated pipette to dilute with 1.0 cm³ of distilled water. Tap the pipette against the inside of the test tube to ensure that all of the solution has been transferred. Label the test tube as 0.9 mol dm⁻³.

3. Repeat step 2 for volumes of 1.0 mol dm⁻³ copper(II) sulfate down to 1.0 cm³ in 1.0 cm³ intervals, diluting each solution up to 10 cm³ with distilled water. Label each solution as: 0.8, 0.7, etc. mol dm⁻³.

Calculations: You need to calculate each volume combination of 1.0 mol dm⁻³ copper(II) sulfate and distilled water.

Expression of results: Write the concentration of copper(II) sulfate solution using the correct units on the side of each test tube.

Exemplar standard procedure

Performing a colorimetric technique to identify the unknown concentration of a sample of copper(II) sulfate

This SP is adapted from *AQA A-level Chemistry*; McFarland, A., Quigg, T. and Henry, N.; Hodder Education 2019.

Scope: This SP can be used to determine the concentration of an unknown concentration of copper(II) sulfate solution with a concentration less than 1.0 mol dm⁻³. This SP could be adapted for any coloured solution.

Principle: When light passes through a coloured solution, some of the light is absorbed by the solution. A calibration curve of absorption against known concentration is produced and used to determine the absorption of a sample with unknown concentration, by comparing the absorption with the calibration curve. A diagram showing the principle of a colorimeter is shown in Figure 2.4.

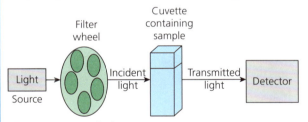

Figure 2.4 A colorimeter

Materials needed: Standard solutions of copper(II) sulfate with a range of different concentrations from 0.1 mol dm⁻³ to 1.0 mol dm⁻³; copper(II) sulfate solution sample with unknown concentration; distilled water.

Equipment needed: Colorimeter; suitable colorimeter filters (or way to vary the wavelength); cuvettes for the colorimeter; pipettes to fill cuvettes; wash bottle for distilled water.

Safety: A risk assessment compatible with your centre should be written or followed. Written risk assessments must be checked by a competent person before use. You should consult the relevant CLEAPSS Student Safety Sheets: 40 Copper and its compounds. Copper(II) sulfate is harmful. Wear eye protection.

Procedure:

1. Turn on the colorimeter and leave for 15 minutes to ensure that the light source is stable and at its correct operating temperature.

2. To avoid cross-contamination, each concentration of solution should have its own pipette and cuvette; label each pipette with the concentration of solution used.

3. Fill a cuvette with 1.0 mol dm⁻³ copper(II) sulfate solution and place into the colorimeter. Select a coloured filter and measure and record the absorbance and the colour of the filter. Repeat for each available filter.

4. Examine the values collected in step 3 and determine which coloured filter results in the greatest absorbance of the light. This is the filter to use for the experiment. (Note: some colorimeters allow you to vary the wavelength of the incident light. You should alter the wavelength until you have identified λ_{max}, the wavelength with maximum absorbance.)

5. Select the chosen filter on the colorimeter. Fill a cuvette with distilled water and place into the colorimeter. Adjust the absorbance reading to zero.

6. Replace the cuvette containing distilled water with the cuvette filled with 1.0 mol dm⁻³ copper(II) sulfate. Measure and record the absorbance.

7. Repeat step 6 with the other concentrations of copper(II) sulfate, ensuring that each one is transferred to a separate clean cuvette using its own pipette to avoid cross-contamination. Measure each absorbance and record your results in a suitable table.

8. Repeat your measurements twice more so that you have three values of absorbance for each concentration of copper(II) sulfate.

9. Identify any anomalous values and repeat the measurements if necessary.

10 Fill a cuvette with the sample of unknown concentration. Insert this cuvette into the colorimeter and measure and record the absorbance.

11 Repeat step 10 twice more so that you have three values (repeat further if any values are anomalous).

Calculations: Calculate the mean average value of absorbance for each concentration and the unknown.

Expression of results: Plot a graph of absorbance against wavelength (if appropriate), or a chart of absorbance against filter colour.

Plot a calibration graph of absorbance against concentration and use this graph to determine the concentration of the unknown sample.

> **Exam tip**
>
> Make sure that you fully reference any researched information that you use in your report.

Applied experimental techniques in physics

Resistivity

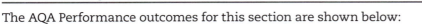

REVISED

The AQA Performance outcomes for this section are shown below:

Performance outcomes	Pass	Merit	Distinction
	To achieve a pass the learner must evidence that they can:	In addition to the pass criteria, to achieve a merit the learner must evidence that they can:	In addition to fulfilling the pass and merit criteria, to achieve a distinction the learner must evidence that they can:
PO3 Demonstrate applied experimental techniques in physics	**P7** Explain the term: resistive to in relation to material properties. (This is assessed in conjunction with the 3(b) Specific Heat Capacity.)	**M7** Describe how the values of resistive to determine the uses of materials in industry. (This is assessed in conjunction with the 3(b) Specific Heat Capacity.)	
PO3(a) Resistivity	**P8** Follow a standard procedure to measure the resistive to of one material and record results.	**M8** Compare results in resistivity with industry standard data, accounting for anomalous readings.	**D5** Compare the methods used in industry to measure the resistivity of materials, including levels of accuracy and precision.
If applicable: **PO4** Understand safety procedure and risk assessment when undertaking scientific practical work. (This is assessed if the Risk assessment has been given to you in 3(b) Specific Heat Capacity.)	**P10** In using experimental techniques: + safely use a range of practical equipment and materials + identify hazards + produce risk assessments for one applied experimental technique from each of biology, chemistry and physics.		

Check your understanding and progress at **www.hoddereducation.co.uk/myrevisionnotes**

Make sure you follow this checklist

In your portfolio you need to cover the following:

☐ Define and explain the meaning of resistivity.
  ☐ Consider the effect of temperature on the resistivity of different types of material.
  ☐ Include any formulae for calculating resistivity; state what the symbols mean and the relevant units involved.

> **Maths skills**
>
> The resistivity of materials varies widely. To compare very large and very small numbers you will have to use standard form. Remember: 156 000 in standard form is 1.56×10^3. 0.00378 is 3.78×10^{-3}. The indices represent powers of 10.

☐ State how the resistance of an electrical component is related to the resistivity of the material that it is made from.
  ☐ Link (mathematically) resistivity and resistance.
  ☐ Include some examples, including metal wires and semiconductors.
  ☐ You could relate the structure of the material to the resistivity. Include diagrams.
☐ State the importance of knowing the resistivity of a material.
  ☐ Link the resistivity of a material to its intended use and state why having a particular value of resistivity determines the various uses of a material (ideally with examples).
☐ Give examples of materials used in the context of your assignment (e.g. those used for components in the automotive industry).
  ☐ Give a range of different examples, including materials with high, low and intermediate values of resistivity, and explain why these materials are used for particular components.
  ☐ Include semiconductors used in electronic circuits.
  ☐ Describe the suitability of these materials for their uses in terms of other, relevant properties, as well as resistivity.
  ☐ Research values of the relevant material resistivities and cross-reference these values with relevant (and reliable) tables of data (such as online material properties databases). Fully reference these tables of data.
☐ Follow or design a risk assessment for this experiment.
  ☐ Your centre will provide you with a risk assessment, or you should use their pro-forma for designing risk assessments. You should have evidence of the risk assessment in your portfolio.
  ☐ If you are designing a risk assessment it should be checked by a competent person (teacher/tutor) before you use it.
  ☐ If you are designing a risk assessment you should consult the relevant CLEAPSS Student Safety Sheets.
  ☐ Check that you have included all the information needed for assessment based on the checklist on page 93.
☐ Follow a given standard procedure (SP) to measure the resistivity of a piece of wire.
  ☐ You may be given a standard procedure by your centre. An exemplar SP for this task is given below. You should include the SP used in your portfolio.
  ☐ Include photos of your experiment set-up and photos taken while the procedure is being carried out.
☐ Record all relevant measurements.
  ☐ Measurements will include:
    ☐ voltage and current readings (if using the voltmeter–ammeter method) or resistances (if using the ohmmeter) method, varied for different lengths of wire over a range
    ☐ measurements of the length and diameter of the wire (a number of different times along the whole length of the wire). When measuring and recording the diameter of the wire it is good practice to also

> **Exam tip**
>
> Beware of confusing resistivity with resistance. They are related to each other but they are not the same thing. We talk about the resistivity of a material but the resistance of a component.

> **Exam tip**
>
> *Define* means that you need to state a definition. *Explain* means that you have to give reasons why scientists need to use the quantity called 'resistivity'.

> **Exam tip**
>
> Choose examples of materials from the context of the brief. If this is the automotive industry, then choose metals, semiconductors and non-metals used in the construction of cars.

record the standard wire gauge (swg) of the wire, and then cross-reference this to your measured values.

☐ Use (and note down) measuring instruments with the relevant precision needed for the experiment.

☐ Repeat all measurements a suitable number of times.

Exemplar standard procedure

Measuring the resistivity of a metal wire

This SP is based on the SP produced by the Institute of Physics TAP project: Episode 112-3 Measuring electrical resistivity.

Scope: This SP can be used to measure the resistivity of a metal wire with a high resistance, such as nichrome or constantan. The wire must be metallic, solid, have no kinks or joins, and must be 1.2 m long.

Principle: The resistivity of a measured area, wire-form sample can be determined experimentally by measuring the resistance of the wire across its length.

Materials needed: 1.2 m of 32 swg (standard wire gauge) constantan or nichrome wire

Equipment needed: Low-voltage (approx. 2 V) dc power supply or battery pack; 2 × crocodile connecting clips; connecting wires; dc voltmeter (0–5 V range, or variable range, e.g. multimeter); dc ammeter (0–10 A range, or variable range, e.g. multimeter); micrometer; metre rule; insulating tape

Safety: A risk assessment compatible with your centre should be written or followed. Written risk assessments must be checked by a competent person before use. Take care with sharp metal wire ends. Wear eye protection.

Procedure:

1 Use a micrometer to measure and record the diameter of the wire, 10 times along the length of the wire.

2 Fix the wire to the metre rule (numbered side) using insulating tape at either end, so that approximately 10 cm of wire is free at each end.

3 Set up the circuit as shown in Figure 2.5. Connect the sample wire to connecting wires using crocodile connecting clips.

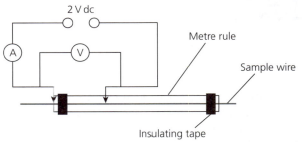

Figure 2.5 Circuit for measuring the resistivity of a wire

4 Fix one crocodile clip to the wire at one end. The other crocodile clip is free to move along the wire.

5 Fix the free crocodile clip to the wire 10 cm from the fixed end.

6 Turn on the power supply and adjust the ranges of the voltmeter and ammeter as required.

7 Measure and record the current, I, and the voltage, V.

8 Turn off the power supply.

9 Repeat steps 5–8 for crocodile clip separations, l, of 20 cm to 100 cm in 10 cm intervals.

10 Repeat steps 5–9 four times, giving five sets of measurements.

Calculations: Calculate the mean average diameter of the wire. Use this value to calculate the cross-sectional area of the wire.

Calculate the mean average values of V and I for each length of wire. Use the mean average values of V and I to calculate mean values of the resistance of the wire, R, for each length.

Expression of results: Express your results by plotting a graph of mean resistance, R against length of wire, l.

☐ Record data with appropriate precision. The data needs to be assessed for its accuracy and reliability.

☐ Identify, deal with and account for any anomalous readings in your data set.

☐ Evaluate the method used to measure the data and try to explain how any anomalies could have occurred.

☐ Calculate a value for the wire's resistivity.

☐ Use the mean values of the data collected to determine the value using the formula for resistivity.

☐ You could use a graphical method to corroborate your value.

☐ State your calculated value to the correct number of significant figures, to match the data that you have collected.

☐ Include the correct units for your calculated value.

☐ Calculate a theoretical percentage error for your calculated value.

☐ To do this, you need to first calculate the percentage error in each of the mean measurements used to calculate the value.

> AQA specify the definitions of the terms accurate, reliable and anomaly. You should refer to these definitions on page 92.

Check your understanding and progress at **www.hoddereducation.co.uk/myrevisionnotes**

- □ The total percentage error is the sum of all the percentage errors of the mean measurements (values that are squared in an equation need to be added twice).
- □ The percentage error should be given to the same number of significant figures as your calculated value.

> **Maths skills**
>
> If a value is calculated from a formula in the form: $y = k\dfrac{ab^2}{c}$; where k is a constant and a, b and c are variables, then the percentage error in y, $\%y = \%a + (2 \times \%b) + \%c$ where $\%a$, $\%b$ and $\%c$ are the percentage errors in a, b and c respectively.

> **Maths skills**
>
> Ensure that all quantities have the correct units. You should generally use SI units where possible. In this case, you are measuring: voltage (in volts, V); current (in amperes, A); length (in cm); and diameter (in mm). You must first convert cm and mm to m: $1\,cm = 1 \times 10^{-2}\,m$ and $1\,mm = 1 \times 10^{-3}\,m$. You can keep measurements in cm and mm, but values such as mean averages must be converted to m before they are used in a formula.

- □ Compare your calculated value with a researched value for the resistivity of the wire's material and give reasons for any difference between theoretical and calculated values.
  - □ Fully reference your researched value.
  - □ Use your theoretical error to assess any difference between your calculated and researched values.
- □ Fully evaluate the methodology of your experiment.
  - □ Evaluate the method of your experiment, i.e. the strengths and weaknesses in your method, and what you could do in future to improve the method.
  - □ As part of the methodology, consider the issues associated with contact resistance, which occurs when electrical components are connected together.
  - □ Generally evaluate the quality of the data obtained. Compare and contrast the percentage errors of each measurement and the precision of the measurements and suggest how the percentage errors could be reduced.
  - □ Cross-reference your values for the diameter of the wire with its stated swg (standard wire gauge).
- □ Compare the methods used in industry to measure resistivity of materials, including levels of accuracy, precision and validity.
  - □ You could consider techniques such as using gold plated connectors, 4 point co-linear probes, Kelvin sensing or the use of a Kelvin bridge.
  - □ You could also consider the different types of resistivity measurement – bulk/volume and sheet resistivity measurements.

> **Exam tip**
>
> Ensure that your researched value is given for the same temperature range as your calculated value.

> AQA specify the definitions of the terms precise, accurate and valid. You should refer to these definitions on page 92.

Specific heat capacity

REVISED ○

The AQA Performance outcomes for this section are shown below:

Performance outcomes	Pass	Merit	Distinction
	To achieve a pass the learner must evidence that they can:	In addition to the pass criteria, to achieve a merit the learner must evidence that they can:	In addition to fulfilling the pass and merit criteria, to achieve a distinction the learner must evidence that they can:

111

Performance outcomes	Pass	Merit	Distinction
PO3 Demonstrate applied experimental techniques in physics	**P7** Explain the term specific heat capacity (SHC) in relation to material properties. (This is assessed in conjunction with 3(a) Resistivity.)	**M7** Describe how the values of SHC determine the uses of materials in industry. (This is assessed in conjunction with 3(a) Resistivity.)	
PO3(b) Specific heat capacity	**P9** Follow a standard procedure to measure the SHC of one material and record results.	**M9** Calculate percentage error and produce a graph to show change in temperature of one material over time and explain the shape of the graph.	**D6** Explain how this standard procedure could be adapted to measure the SHC of a material which is in a different phase.
If applicable: **PO4** Understand safety procedure and risk assessment when undertaking scientific practical work. (This is assessed if the Risk assessment has been given to you in 3(a) Resistivity.)	**P10** In using experimental techniques: + safely use a range of practical equipment and materials + identify hazards + produce risk assessments for one applied experimental technique from each of biology, chemistry and physics.		

Make sure you follow this checklist

In your portfolio you need to complete the following:
- ☐ Define and explain the meaning of specific heat capacity.
  - ☐ Include any formulae for calculating specific heat capacity, state what the symbols mean and define the relevant units.
  - ☐ Give some suitable examples (metals, non-metals, semiconductors and water) from a range of different materials and link the specific heat capacity to the properties of those materials.

Exam tip
Define means that you need to state a definition. *Explain* means that you have to give reasons why scientists need to use the quantity called 'specific heat capacity'.

Maths skills
Some quantities, such as specific heat capacity, have compound units, i.e. their units are made up of two or more other units. In these cases, you should use indices rather than / lines to write the units. An example of one such compound unit is momentum, which has the units $kg\,m\,s^{-1}$ (not $kg\,m/s$).

- ☐ State the importance of knowing the specific heat capacity of a material.
  - ☐ Link the specific heat capacity of a material to its intended use and state why having a particular value of specific heat capacity determines its various uses (ideally with examples).
- ☐ Give examples of materials used in the context of your assignment (e.g. for components in the automotive industry).
  - ☐ Give a range of different examples, including materials with high, low and intermediate values of specific heat capacity, and explain why these materials are used for the intended components.
  - ☐ You must include water.
  - ☐ You should also describe the suitability of these materials for their uses in terms of other relevant properties, as well as specific heat capacity.
  - ☐ Research values of the relevant material specific heat capacities and cross-reference with relevant (and reliable) tables of data (such as online material properties databases). Fully reference these tables of data.

Exam tip
Choose examples of materials from the context of the brief. If this is the automotive industry, then choose metals, semiconductors, non-metals and water, used in the construction and operation of cars.

- ☐ Follow or design a risk assessment for this experiment.
  - ☐ Your centre will provide you with a risk assessment, or you should use their pro-forma for designing risk assessments. You should include a copy of the risk assessment in your portfolio.
  - ☐ If you are designing a risk assessment it should be checked by a competent person (teacher/tutor) before you use it.
  - ☐ Check that you have included all the information needed for assessment based on the checklist on page 93.
- ☐ Follow a given standard procedure (SP) to measure the specific heat capacity of a material in the solid or liquid phase.
  - ☐ You may be given a standard procedure by your centre. An exemplar SP for this task is given below. You need to include a copy of the SP used in your portfolio.
  - ☐ Include photos of your experiment set-up and photos taken while the procedure is being carried out.
  - ☐ You must record all relevant measurements.
    - ☐ This will include (using the SP given below):
      - ☐ the voltage and current readings supplied to the heater
      - ☐ the mass of the metal block
      - ☐ the time
      - ☐ the initial temperature of the block and the temperature of the block every 60 seconds for 10 minutes.
    - ☐ Record the temperature of the block every 60 seconds for 5 minutes after switching off the power supply.
    - ☐ Use (and note down) measuring instruments with the relevant precision needed for the experiment.
    - ☐ Repeat your experiment a suitable number of times.
  - ☐ State the steps taken to minimise thermal heat loss from your experiment.

> ### Exam tip
>
> Most students choose to perform an experiment to determine the specific heat capacity of a *metal block*, designed specifically for the purpose of this experiment. This is a good idea.

Exemplar standard procedure

Measuring the specific heat capacity of a metal block

This SP is based on the SP produced by the Institute of Physics Practical physics project: Specific thermal capacity of aluminium more accurately.

Scope: This SP can be used to measure the specific heat capacity of a metal block, such as aluminium, steel, brass or copper. The block must be shaped to include holes for a standard 12 V dc laboratory heater and a thermometer.

Principle: The specific heat capacity of a metal block can be determined experimentally by measuring the thermal energy absorbed by a measured mass of block for a measured temperature change.

Materials needed: Shaped metal block (aluminium, steel, brass or copper)

Equipment needed: 12 V dc power supply or battery pack; connecting wires; dc voltmeter (0–20 V range, or variable range, e.g. multimeter); dc ammeter (0–10 A range, or variable range, e.g. multimeter); electronic balance capable of measuring approx. 1 kg; 12 V dc heater (to fit hole in the block); thermometer (to fit hole in the block); stopwatch; insulating material (e.g. bubble wrap) cut to fit the block; heatproof mat

Safety: A risk assessment compatible with your centre should be written or followed. Written risk assessments must be checked by a competent person before use. Take care with hot apparatus. Wear eye protection.

Procedure:

1. Record the material that the block is made from.
2. Measure and record the mass of the metal block. Repeat the measurement twice giving three values in total.
3. Set up the apparatus and circuit as shown in Figure 2.6.
4. Once set up, allow the apparatus to come to thermal equilibrium by waiting for one minute. Then record the starting temperature.

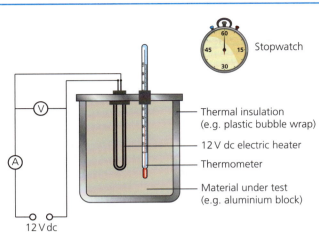

Stopwatch

Thermal insulation (e.g. plastic bubble wrap)

12 V dc electric heater

Thermometer

Material under test (e.g. aluminium block)

12 V dc

Figure 2.6 Experiment to measure the specific heat capacity of a metal block

5 Adjust the output voltage of the power supply unit to 12 V and switch on. Start the stopwatch.

6 Measure and record the temperature, T; current, I; and voltage, V.

7 Repeat step 6, every 60 s for 10 minutes (600 s).

8 Switch off the power supply, but keep recording the temperature of the block every 60 s for 5 minutes (300 s).

9 Allow the whole apparatus to cool back to room temperature.

10 Repeat steps 5–9 twice more, giving three sets of measurements.

Calculations: Calculate the mean mass of the block.

Calculate the mean average values of V and I. Use these values to calculate the mean average power of the 12 V heater.

Calculate the temperature change for each time, t, and then use these values to calculate the mean temperature change, ΔT, for each time.

Expression of results: Express the results by plotting a graph of mean temperature change, ΔT, against time, t (in seconds).

☐ Record data with appropriate precision. Assess the accuracy and reliability of your data.

☐ Identify, deal with and account for any anomalous readings in your dataset.

  ☐ Evaluate the method used to measure the data and try to explain how any anomalies could have occurred.

☐ Calculate a value for the specific heat capacity of the material from which the block is made.

  ☐ Plot a graph of mean change in temperature against time. An annotated exemplar version of this is shown in Figure 2.7.

  ☐ Determine and plot error bars for each point on your graph using the uncertainties in the temperature measurements and the spread of the values.

  ☐ Give reasons for and explain the shape of the graph.

  ☐ Use the linear portion of the graph, and the error bars, to determine the best-fit temperature change over the linear heated section of the graph, together with an error value.

  ☐ Calculate the thermal energy supplied to the block in the linear heated section of the graph using the mean power of the heater and the time during the linear section.

  ☐ Calculate a final value for the specific heat capacity of the material of the block using thermal energy supplied, the mass of the block and the best fit temperature change in the linear heated portion of the graph.

  ☐ State your calculated value to the correct number of significant figures, to match the data that you have collected.

  ☐ Include the correct units for your calculated value.

☐ Calculate a theoretical percentage error for your calculated value.

  ☐ To do this, you need to first calculate the percentage error in each of the mean measurements used to calculate the specific heat capacity. This includes:

    ☐ the percentage error in the mean mass, determined using the spread of the measured masses and the uncertainty of the balance

    ☐ the percentage error in the thermal energy supplied to the heater in the linear heated portion of the graph. You can do this using the spread of the error bars on your graph. This should give you a maximum and minimum value for the temperature, from which you can calculate the percentage error.

AQA specify the definitions of the terms precise, accurate, anomaly and reliable. You should refer to these definitions on page 92.

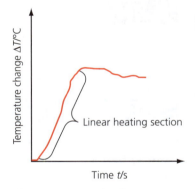

Linear heating section

Time t/s

Figure 2.7 Graph of mean temperature change against time

Exam tip

Remember to use seconds for your graph, and seconds in your calculation of the thermal energy supplied by the heater to the metal block.

Check your understanding and progress at **www.hoddereducation.co.uk/myrevisionnotes**

- □ The total percentage error is the sum of all the percentage errors of the mean measurements.
- □ The percentage error should be given to the same number of significant figures as your calculated value.
- □ Explain how the shape of the heating/cooling graph affects the validity of the calculated value of the specific heat capacity.

- □ Compare your calculated value with a researched value for the specific heat capacity of the block's material and give reasons for any differences between theoretical and calculated values.
  - □ Fully reference your researched value.
  - □ Use your theoretical error to assess any difference between your calculated and researched values.
  - □ Calculate a percentage error of your calculated value compared with the theoretical researched data book value.
  - □ Suggest reasons for any differences between your calculated value and the theoretical researched data book value.
- □ Fully evaluate the methodology of your experiment.
  - □ Evaluate the method of your experiment, i.e. the strengths and weaknesses in your method, and what you could do in future to improve the method.
  - □ As part of the methodology, consider the issues associated with thermal heat transfer and thermal heat loss/cooling effects, which occur when there is a difference in temperature between the heated apparatus and the surroundings. This could involve explanations of the non-linear sections of the graph. You could also research Newton's Law of cooling and discuss how this could be applied to your experiment.
  - □ Generally evaluate the quality of the data obtained. Compare and contrast the percentage errors of each measurement and the precision of the measurements and suggest how the percentage errors could be reduced.
- □ Compare the methods used in industry to measure the specific heat capacity of materials, including levels of accuracy, precision and validity.
  - □ You could consider techniques such as differential scanning calorimetry (DSC) and the transient hot bridge (THB) method.
  - □ You could also consider how specific heat capacity varies with temperature and how it varies with phase (solid, liquid or gas).
- □ Describe how the method could be adapted for a material in a different phase.
  - □ If you have determined the specific heat capacity of a solid for the main part of your experiment, you should describe how the method would change for a material in the liquid phase, such as water (or vice versa).
  - □ Consider any extra precautions that you would need to take to ensure that the results from your modified experiment were still accurate and valid.

> AQA specify the definitions of the terms precise, accurate and valid. You should refer to these definitions on page 92.

Exam tip

You do not need to perform an experiment to measure the specific heat capacity of a material in a different phase.

Unit 3 Science in the modern world

Introduction

Unit 3 will develop your skills at interpreting how science is used in the modern world and how it is perceived by the media. You will need to:
+ interpret information
+ process and present data
+ evaluate the usefulness and appropriateness of information.

You will do this by:
+ reading and interpreting a variety of scientific texts
+ engaging with topical scientific issues
+ discussing the ethical and social implications of scientific issues
+ exploring how science is represented in the media
+ exploring how scientists work and the roles that they carry out.

This unit is assessed via an external examination, set by AQA. There are two sections to the examination.
+ Section A consists of questions based on a pre-released set of articles about a common theme. The pre-released articles are published at the end of March for the June examination series and the beginning of November for the January examination series.
+ Section B consists (generally) of data analysis questions based on tabulated sets of data in the examination paper that may or may not be loosely related to the context of the pre-release, along with (usually) questions about the roles and responsibilities of scientific personnel.

> **Exam tip**
>
> Read the questions very carefully. In particular, make sure that you identify the question *command* word. A full glossary of the terms used in Applied Science examinations is available on the AQA website. The most common mistake made by candidates is to describe something when the question asks for an evaluation. If a question asks you to *evaluate* something, you need to consider its *strengths and weaknesses*.

Dealing with the pre-release document REVISED

You must write *succinctly*; this is most important for the extended response questions. As a general rule, unless your handwriting is really big, try to limit the extent of your answers to the number of lines provided.

You should regularly review the pre-released source booklet between the release date and your examination. It is suggested that you highlight and annotate the booklet as required. You will be given a new copy for the examination. You cannot take your notes into the examination.

A good technique for studying the pre-release material is to briefly summarise the key points of each source as bullet points at the end of each article. You can then use these points for immediate revision, just before the examination.

It is not the purpose of this unit to go through every possible scientific scenario that could be put into the pre-release. Instead, this unit will give you a checklist of things that you need to address when going through the pre-release with your teacher/tutor in your centre.

Previous pre-release contexts have been:
+ fracking
+ heart attacks
+ micro-plastics
+ the Chernobyl accident
+ genetic modification
+ electric cars.

The pre-release document will contain a selection of text-based sources about a common theme. There will typically be about four or five sources, labelled A, B, C, D, etc. One or two of the sources may include diagrams, pictures or charts.

It is important to realise that, due to copyright restrictions, AQA will refer to the sources used in the pre-release as 'Adapted Articles'. This does not mean that the article has been changed in any way, and it *does not affect the validity of the source* of the article. You should ignore the fact that it has been 'adapted' and focus instead on the origin of the source and its date. You may be given a link to the original article.

Exam tip

You are not expected to carry out any wider reading around the scientific context of the pre-released materials.

Assessment objective 1: Topical scientific issues obtained from a variety of media sources

REVISED

+ The sources given to you in the pre-release will represent a range of articles about a common theme. The theme will be topical and will deal with scientific issues and ideas.
+ The first source is usually from a reputable popular scientific source that gives the background to the issues dealt with in the context. It frequently gives a factual account of the context.
+ The subsequent sources will be from a mix of different sources. They may:
  + express opposite views
  + represent sensational views from the tabloid press or the internet
  + give case studies or first-hand reports of how individuals have been affected by the context.
+ You will need to assess the scientific credentials of the sources for bias or false or sensationalist claims. There are several different categories of source:
  + Specialist scientific publishers, such as universities, scientific journals and textbooks, and popular scientific magazines such as *New Scientist*, *Nature* and *Scientific American*. These are likely to give reliable and unbiased information. Factual information in these publications is usually peer-reviewed.
  + Blogs written by scientists or specialist scientific journalists or experts. These could be reliable but the writers may be writing on behalf of, or in support of, a particular organisation, so may be subject to bias. They are rarely peer-reviewed.
  + 'Broadsheet' newspapers, such as *The Guardian*, *The Times* or the *Daily Telegraph*. Articles written for these publications are generally written by journalists or specialist journalists, who may not be scientifically trained. Sometimes newspapers are subject to bias.
  + Tabloid newspapers, such as the *Sun*, the *Daily Mail*, the *Mirror*, etc. These publications are generally interested in sensationalist reporting. They are likely to be written by general journalists, may not be reliable, and may be subject to bias.
  + Popular blogs written by bloggers or activists. These are generally biased and may be selective with scientific facts and opinions.
+ Be aware of the main organisations behind scientific issues and campaigns. Examples include:
  + The Royal Society
  + Institute of Physics; Royal Society of Biology; Royal Society of Chemistry
  + Greenpeace
  + Campaign for Science and Engineering (CaSE)
  + Women in Science and Engineering (WISE)
  + Friends of the Earth.

Bias A source may show a bias where it shows an inclination or a prejudice for or against a viewpoint, person or group.

Sensationalist A source is said to be sensationalist if it presents a scientific story in a way that is intended to provoke public interest or excitement at the expense of scientific accuracy.

117

+ You will be required to interpret both textual and numerical information within the sources and show a clear understanding of the context. This could include:
    + simple comprehension of the text of each source, for example, being asked to find information within the text
    + giving an explanation based on text within the sources
    + the extraction of numerical information from a source.
+ You will also be asked to process data from the sources, for example, by:
    + calculating simple sums or differences
    + calculating a ratio
    + calculating percentages from data
    + calculating percentage increases of data
    + substituting data into a given formula.
+ You will need to make judgements about the usefulness and appropriateness of the data given. You may need to consider the following:
    + If there is conflicting data, which set do you use?
    + Is the data presented in the source actually relevant to the question?
    + Will the data give an answer to the question?
+ You may be required to interpret and present data within the sources in different formats, including:
    + graphs
    + charts
    + diagrams
    + tables.

You need to practise interpreting these sorts of graphics and selecting and processing data from them. Your teacher/tutor can provide you with plenty of examples, or you could search for examples on the internet.

> **Exam tip**
>
> You need to practise doing straightforward calculations. Your teacher/tutor may provide you with examples, but you can always find sources containing scientific data on the internet. 'Popular science' is a good key term to use in a search engine to find data to use.

Now test yourself

TESTED ◯

1 Describe what is meant by a source.
2 List **four** different categories of source.
3 Explain why an article in a scientific journal is likely to be more valid than an article about the same topic in a tabloid newspaper.
4 The mission of the Royal Meteorological Society (RMS) is to 'engage, enthuse and educate by promoting the understanding and application of weather and climate science for the benefit of all'. The mission statement of Friends of the Earth (FotE) is 'to collectively ensure environmental and social justice, human dignity, and respect for human rights and peoples, so as to secure sustainable societies'. Suggest and explain a difference between the way that these two organisations might react to the observation that the Greenland ice sheet shrank by a record 532 billion tonnes in 2019, compared to the annual average of 255 billion tonnes since 2003.
5 Using the data from Question 4:
  a calculate the increase in ice sheet shrinkage (in billion tonnes) in 2019, compared to the annual average
  b calculate the percentage increase in the loss of ice.

Assessment objective 2: The public perception of science and the influence that the media have

REVISED ◯

Make sure you follow this checklist

You need to be able to do the following:
☐ Consider how science is developed and the ways that it is used when communicating with different types of audience.
  ☐ Good science requires the clear use of scientific method. It also requires good peer review and corroboration.

> **Corroboration** Using evidence which confirms or supports a scientific statement, theory or finding.

Check your understanding and progress at **www.hoddereducation.co.uk/myrevisionnotes**

- □ Scientific work is expensive and requires funding. Such funding is mostly provided by governments (through research councils in the UK), charities or commercial companies (particularly in the case of pharmaceuticals). The main scientific institutions are universities, specialist research laboratories (such as those run by the Science and Technology Facilities Council or the Wellcome Trust) and companies (such as GlaxoSmithKline and AstraZeneca).
  - □ Scientists communicate their work mainly through the medium of scientific papers, published in scientific journals (usually online). These papers may then be referenced by other forms of media, who use scientific journalists to interpret the work for other audiences.
- □ Consider the different approaches and styles used by the media when communicating with scientists and with wider society.
  - □ You may need to consider differences in the ways that the general and specialist media report on scientific issues.
  - □ Depending on the audience type, the media use different tones and language to describe scientific issues. If the audience is considered to be scientists or relatively scientifically literate, then more scientific language will be used and the tone will be more matter of fact. Other, less scientific publications tend to be more sensationalist and use fewer scientific terms.
  - □ In the examination, you may be asked to identify the target audience for a source. To do this, you must assess the language used, both in the title/strap line and in the main text. Look for clues in the text such as the number of scientific words used or the level of the grammar. Imagine people you know, of different ages and backgrounds, and think how they would read and react to the source. You may even be asked to give examples/quotes from the text to corroborate your ideas.
  - □ The popular and sensationalist media's reporting of scientific issues is often over-simplified and sometimes inaccurate; the reported science may be carefully selected to make a point. Remember that non-scientific publications do not have to be peer-reviewed, so may not have the same validity. This has become part of what is now known as 'fake news'.
- □ Know the ways that scientists publish their work and share it with others. You must be able to explain the process of peer review.
  - □ The peer review process consists of the following steps:
    - □ A scientist or researcher submits an article/paper to a scientific journal.
    - □ The journal sends the article/paper to an anonymous reviewer who is qualified in the same scientific field.
    - □ The reviewer comments on or checks the article/paper.
    - □ The commented/corrected article/paper is amended by the scientist/researcher or the journal approves the article/paper without changes.
    - □ The whole cycle is repeated as necessary
  - □ Peer review ensures that new scientific discoveries or advances are checked and corroborated by independent scientists who are experts in the same field. The genuine scientific community is used to this process and there are clear ways for scientists to quickly share their data and ideas via 'pre-print' scientific servers, forums and blogs before the peer-review process is carried out. This is particularly important in the case of major discoveries such as new treatments for diseases.
  - □ Be prepared to identify sources that have not been peer-reviewed and are therefore less valid than sources that have been peer-reviewed.
- □ Be able to describe the ways in which scientists gather data, hypotheses, arguments and theories, and describe the ways in which they are used.
  - □ Scientific method consists of the following steps:
    - □ observation (of a phenomenon)
    - □ question (identify the problem)
    - □ research (search for existing solutions)

- □ hypothesise (come up with a hypothesis)
- □ experiment (design and carry out an experiment to collect data)
- □ test (analyse the data to accept or reject the hypothesis)
- □ conclusion (make a conclusion related to the hypothesis)
- □ report (report results).
- □ Scientists can gather data in a variety of ways:
  - □ directly from laboratory experiments or specialist observatories
  - □ from experiments set up in the field and monitored *in situ* or remotely
  - □ by direct observation in the field
  - □ by population testing (e.g. double-blind trials)
  - □ via remote sensing from altitude or by satellites
  - □ remotely by population tagging.
- □ A scientific hypothesis is a prediction that can be tested by experiment. Hypotheses are generally formulated from previous observations or data, or by new theories.
- □ We define a scientific argument as the process that scientists go through when they disagree about scientific explanations (or claims). Scientists use empirical data (or evidence) to justify their side of the argument (or rationale). This process can guide the work of scientists who identify weaknesses and limitations in other scientists' arguments, with a view to refining and improving scientific explanations and the way that experiments are designed. This process is known as *evidence-based argumentation*.
- □ A scientific theory explains why a scientific phenomenon occurs and allows different scientists to produce a range of scientific hypotheses.
- □ Consider the ways that society and the media interact with science, and the fact that the media give scientists a platform for explaining their work, in turn giving the public a way to understand science.
  - □ It is important to remember that the media give scientists a platform to communicate their work to wider audiences. This can, however, give the media the power to set the scientific agenda and influence opinion on scientific issues in ways that are not always consistent with the science.
  - □ You need to be aware that the popular media can often hijack scientific stories and this can lead to the notion of 'bad science' or 'fake science'.

Now test yourself

TESTED

6 Describe the process of scientific method.

7 A scientist is working in a university, researching a new vaccine. State the organisations that they could apply to for funding.

8 Describe the process of peer-review.

9 The first 'approved' vaccine for Covid-19 was produced in Russia. The team responsible did not release their trial data for peer review. Suggest possible consequences of this decision.

10 A scientist is interested in researching the increase of locust infestations in East Africa. Suggest **two** ways that they could do this.

11 Describe the process of evidence-based argumentation.

12 In 1998, Dr Andrew Wakefield published an article in *The Lancet*, proposing a link between the MMR vaccine injection and autism. His theory was based on a 12-patient case study, but was picked up by the mainstream media; as a result, the number of children in the general population being vaccinated dropped substantially. Explain why this is an example of 'bad science' and why this is an example of the poor use of science by the media.

Assessment objective 3: The ethical, moral, commercial, environmental, political and social issues involved in scientific advances, and how these are represented in the media

Make sure you follow this checklist

You need the following skills:

- ☐ Know how the benefits and drawbacks of topical scientific advances are presented in the media (e.g. how the media treat subjects such as GM crops or fracking).
  - ☐ The mainstream media are quick to pick up on topical scientific advances. However, because they are not subject to the peer-review process, poor science can often be sensationalised.
  - ☐ The benefits and drawbacks (in other words, a balanced view) are not always regarded as a good story, so one aspect is often overlooked.
- ☐ Be able to assess any environmental, commercial or health and safety implications of scientific advances.
  - ☐ Scientific advances usually have benefits and drawbacks. Nuclear power creates large quantities of electricity in a way that is reliable and does not produce greenhouse gases during generation. The drawbacks, however, present huge environmental, commercial and health and safety issues.
  - ☐ In many cases the scientific advance should be carefully considered – do the benefits outweigh the drawbacks? With nuclear power, judgements must be made by government as to the overall merit of a project.
- ☐ Be able to identify any social, ethical and moral matters associated with scientific advances.
  - ☐ Do not give vague/sweeping answers to questions that ask you to identify any issues. This is particularly important for moral issues.
  - ☐ Common examples of this include animal research, drug trials and organ transplantation.
- ☐ Know how the media treat any social, ethical and moral issues associated with scientific advances.
  - ☐ You need to be able to pick out relevant examples from the sources provided. You may be asked to identify an issue and then assess how the author(s) of the source present the issue. Is the issue presented in a positive or negative way?
  - ☐ Different categories of media will present these issues in different ways.
- ☐ Be able to assess how scientific advancements are influenced by national and/or local political pressure groups.
  - ☐ National pressure groups, such as Greenpeace, look to put pressure on government, to get them to address key issues.

> ### Exam tip
>
> Avoid using statements such as 'playing god' or 'not natural' in your answers – they are not specific enough.

Now test yourself

 TESTED ○

13 The UK Government approved the construction of the £22.9 billion Hinkley Point C nuclear reactor, near Bristol, in 2012, and construction began in 2018. Describe **two** benefits and **two** drawbacks for the UK population of this project.

14 One of the arguments for nuclear power is that it provides reliable, uninterrupted, large quantities of mains electricity. Suggest **one** social issue associated with having an unreliable electricity supply.

15 Hydroxychloroquine is a common, cheap and widely available drug used to treat malaria. At the beginning of the Covid-19 epidemic, several high-profile media-savvy personalities and politicians began to publicise a very small-scale study in France that suggested hydroxychloroquine could help to treat coronavirus patients. Many larger-scale studies have since disproved this. Describe **two** ways in which the media has reacted to this scientific story.

16 Greenpeace's mission statement says 'Greenpeace is an independent, non-profit, global campaigning organisation that uses non-violent, **creative confrontation** to expose global environmental problems and their causes. Greenpeace's goal is to ensure the ability of Earth to nurture life in all its diversity'. Suggest a way that Greenpeace might use 'creative confrontation' to influence government policy on environmental problems.

Assessment objective 4: The roles and responsibilities that science personnel carry out in the science industry

REVISED ⬤

Make sure you follow this checklist

You need to:

- ☐ Know the different roles of a range of scientists in an organisation, including:
  - ☐ biologist (including marine and zoologist)
  - ☐ biomedical scientist, including microbiologist
  - ☐ chemist, including biochemist and analyst
  - ☐ environmental scientist (ecologist)
  - ☐ geneticist
  - ☐ material scientist
  - ☐ pharmacologist
  - ☐ physicist
  - ☐ product/process developer or technologist, e.g. polymers or food (biotechnologist)
  - ☐ radiographer/radiologist
  - ☐ research scientist
  - ☐ scientific laboratory technician
  - ☐ sport and exercise scientist
  - ☐ toxicologist.
- ☐ Be aware of the scientific skills, techniques and experience that different roles and responsibilities need within an organisation.
- ☐ Know the roles and responsibilities associated with key scientific personnel within an organisation.
  - ☐ Consider organisations that manufacture or process scientific products *and* those that provide a scientific service.
- ☐ Know the benefits of scientific roles to society.
- ☐ Be aware of the relationships between the different scientific personnel in an organisation.
  - ☐ This means that you need to identify other roles of scientific personnel that scientists need to work with. For example, in a hospital, it is likely that pharmacologists will work closely with biomedical scientists.

> **Examiner tip**
>
> When describing the roles of scientists, avoid using vague terms such as 'find out' or 'look at' or 'study'. Instead, use more precise terms such as 'test' or 'conduct experiments' or 'monitor'.

There is nearly always a question about the roles of two or three different scientific personnel on every Unit 3 examination paper. They usually restrict the choices to the list above. You need to learn the roles and responsibilities of each one. You could complete a table such as the example below. Learn the table.

Scientific role	Responsibilities	Scientific skills, techniques and experience needed	Product manufacture/ processing OR scientific service	Benefits of the role to society	Relationships with other key scientific personnel
Biologist					

Check your understanding and progress at **www.hoddereducation.co.uk/myrevisionnotes**

TESTED

Now test yourself

17 A cricket equipment manufacturer is interested in the use of graphene in the design and manufacture of cricket helmets, which need to be light but very strong. Suggest **one** role for each of the following scientists involved with the process of designing and making a new graphene-based cricket helmet.

  a Materials scientist

  b Product/process developer

  c Sport and exercise scientist

18 The Environment Agency is involved with monitoring rivers when there are suspected cases of pollution. Suggest **one** role for each of the following scientists involved with the process of monitoring a river following a suspected pollution incident.

  a Biologist (freshwater zoologist)

  b Ecologist

  c Toxicologist

Summary

Topical scientific issues obtained from a variety of media sources

+ Identify the scientific ideas in topical scientific issues.
+ Interpret textual and numerical scientific information from the media and show a clear understanding of the content.
+ Process data taken from the media and assess it for usefulness and appropriateness.
+ Present data in an appropriate form.

The public perception of science and the influence that the media have

+ Scientific knowledge is developed using 'scientific method' and communicated in different ways with different audiences.
+ The media use different approaches and styles when communicating with scientists and wider society.
+ Scientists use peer review when publishing and sharing their work.
+ Scientists gain and use data, hypothesis, argument and theory in different ways.
+ The media give scientists a platform for explaining their work and provide the public with a way to understand key scientific features, thus allowing society and the media to interact with science.

The ethical, moral, commercial, environmental, political and social issues involved in scientific advances, and how these are represented in the media

+ Scientific advances nearly always come with benefits and drawbacks for different groups of people and the media often represent the advances in different ways depending on the audience.
+ Scientists, and society in general, must consider the environmental and commercial considerations associated with scientific advances, particularly any health and safety implications.

+ Some scientific advances, such as animal research or drug trials, present particular social, ethical and moral matters which must be considered by society.
+ The media treat and present social, ethical and moral scientific issues in different ways.
+ National and/or local political pressure groups play an important role in influencing scientific advancements.

The roles and responsibilities that science personnel carry out in the science industry

+ You should know about the varied roles that scientists can perform in an organisation, including:
  + biologist (including marine and zoologist)
  + biomedical scientist, including microbiologist
  + chemist, including biochemist and analyst
  + environmental scientist (ecologist)
  + geneticist
  + materials scientist
  + pharmacologist
  + physicist
  + product/process developer or technologist, e.g. polymers or food (biotechnologist)
  + radiographer/radiologist
  + research scientist
  + scientific laboratory technician
  + sport and exercise scientist
  + toxicologist.
+ To undertake scientific specific roles and responsibilities within an organisation, some staff need scientifically related skills, techniques and experience.
+ You need to be aware of the roles and responsibilities associated with science personnel within an organisation.
+ People in scientific roles contribute benefits to society.
+ The scientific personnel in an organisation collaborate with each other.

Exam practice

The following questions have been adapted from the AQA L3 Applied Science Unit 3 Science in the Modern World examination June 2018. The original Pre-release Material contained **four** sources of information – these example questions are based on **two** of the Sources (A and B) and some of the Section B data-based questions.

AQA Pre-release Material June 2018

Source A: Adapted article from World Nuclear Association website, November 2016

Chernobyl Accident 1986

(Updated April 2020)

- The Chernobyl accident in 1986 was the result of a flawed reactor design that was operated by inadequately trained personnel.
- The resulting steam explosion and fires released at least 5% of the radioactive reactor core into the environment, with the deposition of radioactive materials in many parts of Europe.
- Two Chernobyl plant workers died due to the explosion on the night of the accident, and a further 28 people died within a few weeks as a result of acute radiation syndrome.
- The United Nations Scientific Committee on the Effects of Atomic Radiation has concluded that, apart from some 6500 thyroid cancers (resulting in 15 fatalities), 'there is no evidence of a major public health impact attributable to radiation exposure 20 years after the accident.'
- Some 350,000 people were evacuated as a result of the accident, but resettlement of areas from which people were relocated is ongoing.

On 25 April, prior to a routine shutdown, the reactor crew at Chernobyl 4 began preparing for a test to determine how long turbines would spin and supply power to the main circulating pumps following a loss of main electrical power supply. This test had been carried out at Chernobyl the previous year, but the power from the turbine ran down too rapidly so new voltage regulator designs were to be tested.

A series of operator actions, including the disabling of automatic shutdown mechanisms, preceded the attempted test early on 26 April. By the time the operator moved to shut down the reactor, the reactor was in an extremely unstable condition. A peculiarity of the design of the control rods caused a dramatic power surge as they were inserted into the reactor.

The interaction of very hot fuel with the cooling water led to fuel fragmentation along with rapid steam production and an increase in pressure. The design characteristics of the reactor were such that substantial damage to even three or four fuel assemblies would – and did – result in the destruction of the reactor. The overpressure caused the 1000t cover plate of the reactor to become partially detached, rupturing the fuel channels and jamming all the control rods, which by that time were only halfway down. Intense steam generation then spread throughout the whole core (fed by water dumped into the core due to the rupture of the emergency cooling circuit) causing a steam explosion and releasing fission products to the atmosphere. About 2–3 seconds later, a second explosion threw out fragments from the fuel channels and hot graphite. There is some dispute among experts about the character of this second explosion, but it is likely to have been caused by the production of hydrogen from zirconium-steam reactions.

Two workers died as a result of these explosions. The graphite (about a quarter of the 1200 tonnes of it was estimated to have been ejected) and fuel became incandescent and started a number of fires, causing the main release of radioactivity into the environment. A total of about 14 EBq (14×10^{18} Bq) of radioactivity was released, over half of it being from biologically-inert noble gases.

About 200–300 tonnes of water per hour was injected into the intact half of the reactor using the auxiliary feedwater pumps but this was stopped after half a day owing to the danger of it flowing into and flooding units 1 and 2. From the second to the tenth day after the accident, some 5000 tonnes of boron, dolomite, sand, clay and lead were dropped onto the burning core by helicopter in an effort to extinguish the blaze and limit the release of radioactive particles.

Immediate impact of the Chernobyl accident

The accident caused the largest uncontrolled radioactive release into the environment ever recorded for any civilian operation, and large quantities of radioactive substances were released into the air for about 10 days. This caused serious social and economic disruption for large populations in Belarus, Russia and Ukraine. Two radionuclides, the short-lived iodine-131 and the long-lived caesium-137, were particularly significant for the radiation dose they delivered to members of the public.

It is estimated that all of the xenon gas, about half of the iodine and caesium, and at least 5% of the remaining radioactive material in the Chernobyl 4 reactor core (which had 192 tonnes of fuel) was released in the accident. Most of the released material was deposited close by as dust and debris, but the lighter material was carried by wind over Ukraine, Belarus, Russia, and to some extent over Scandinavia and Europe.

The casualties included firefighters who attended the initial fires on the roof of the turbine building. All these fires were put out in a few hours, but radiation doses on the first day caused 28 deaths – six of which were firemen – by the end of July 1986.

The next task was cleaning up the radioactivity at the site so that the remaining three reactors could be restarted, and the damaged reactor shielded more permanently. About 200000 people ('liquidators') from all over the Soviet Union were involved in the recovery and clean-up during 1986 and 1987. They received high doses of radiation, averaging around 100 millisieverts (mSv). Some 20000 liquidators received about 250 mSv, with a few receiving approximately 500 mSv. Later, the number of liquidators swelled to over 600000, but most of these received only low radiation doses. The highest doses were received by about 1000 emergency workers and onsite personnel during the first day of the accident.

Initial radiation exposure in contaminated areas was due to short-lived iodine-131; later caesium-137 was the main hazard. (Both are fission products dispersed from the reactor core, with half lives of 8 days and 30 years, respectively. 1.8 EBq of I-131 and 0.085 EBq of Cs-137 were released.) About five million people lived in contaminated areas of Belarus, Russia and Ukraine (above 37 kBq/m² Cs-137 in soil) and about 400000 lived in more contaminated areas of strict control by authorities (above 555 kBq/m² Cs-137). A total of 29 400 km² was contaminated above 180 kBq/m².

The plant operators' town of Pripyat was evacuated on 27 April (45000 residents). By 14 May, some 116000 people who had been living within a 30-kilometre radius had been evacuated and later relocated. About 1000 of these returned unofficially to live within the contaminated zone. Most of those evacuated received radiation doses of less than 50 mSv, although a few received 100 mSv or more.

In the years following the accident, a further 220000 people were resettled into less contaminated areas, and the initial 30 km radius exclusion zone (2800 km²) was modified and extended to cover 4300 square kilometres. This resettlement was due to application of a criterion of 350 mSv projected lifetime radiation dose, though in fact radiation in most of the affected area (apart from half a square kilometre close to the reactor) fell rapidly so that average doses were less than 50% above normal background of 2.5 mSv/yr.

Long-term health effects of the Chernobyl accident

Several organisations have reported on the impacts of the Chernobyl accident, but all have had problems assessing the significance of their observations because of the lack of reliable public health information before 1986.

In 1989, the World Health Organization (WHO) first raised concerns that local medical scientists had incorrectly attributed various biological and health effects to radiation exposure. Following this, the Government of the USSR asked the International Atomic Energy Agency (IAEA) to coordinate an international experts' assessment of the accident's radiological, environmental and health consequences in selected towns of the most heavily contaminated areas in Belarus, Russia and Ukraine. Between March 1990 and June 1991, a total of 50 field missions were conducted by 200 experts from 25 countries (including the USSR), seven organisations and 11 laboratories. In the absence of pre-1986 data, it compared a control population with those exposed to radiation. Significant health disorders were evident in both control and exposed groups but, at that stage, none was radiation related.

Main environmental pathways of human radiation exposure

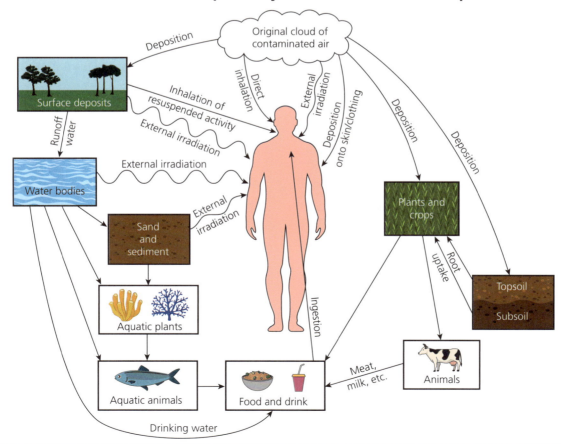

In February 2003, the IAEA established the Chernobyl Forum, in cooperation with seven other UN organisations as well as the competent authorities of Belarus, the Russian Federation and Ukraine. In April 2005, the reports prepared by two expert groups – 'Environment', coordinated by the IAEA, and 'Health', coordinated by the WHO – were intensively discussed by the Forum and eventually approved by consensus. The conclusions of this 2005 Chernobyl Forum study (revised version published 2006) are in line with earlier expert studies, notably the UNSCEAR 2000 report which said that: 'apart from this [thyroid cancer] increase, there is no evidence of a major public health impact attributable to radiation exposure 14 years after the accident. There is no scientific evidence of increases in overall cancer incidence or mortality or in non-malignant disorders that could be related to radiation exposure.'

There is little evidence of any increase in leukaemia, even among clean-up workers where it might be most expected. Radiation-induced leukaemia has a latency period of 5–7 years, so any potential leukaemic cases due to the accident would already have developed. A low number of the clean-up workers, who received the highest doses, may have a slightly increased risk of developing solid cancers in the long term. To date, however, there is no evidence of any such cancers having developed. Apart from these, the United Nations Scientific Committee on the Effects of Atomic Radiation (UNSCEAR) said: 'The great majority of the population is not likely to experience serious health consequences as a result of radiation from the Chernobyl accident. Many other health problems have been noted in the populations that are not related to radiation exposure.'

Source B: Adapted article from Ripley's website, April 2016

CHERNOBYL'S DEADLY ELEPHANT'S FOOT

There's a structure at the heart of the Chernobyl Power Plant known as the elephant's foot, and it can kill you in 300 seconds.

Ripley's Believe It or Not!®
April 29, 2016

THE CHERNOBYL DISASTER

On April 26, 1986, during a routine test, the Number 4 reactor at the Chernobyl Nuclear Power Plant had a power surge and triggered an emergency shutdown. Instead of shutting down, the reactor kept surging power, and in no time at all the plant was in full disaster mode. The control rods used to manage the core's temperature were inserted too late into the process. Instead of cooling down, the rods cracked in the rising heat from the core and locked into place.

As if that wasn't bad enough, the water used to cool the entire reactor vaporised, resulting in a massive explosion. The first explosion blew the lid of the reactor through the roof of the building. The second explosion followed shortly thereafter and sent broken core material, fire, and radioactive waste into the air.

Without the tons of steel and concrete typically used to shield it, the core of the reactor began to melt. The result of the melting process is a substance called corium. Corium is a lava-like molten mixture of portions of the nuclear reactor core, nuclear fuel, fission products, and control rods. At Chernobyl, the corium melted through the bottom of the reactor vessel, oozed through pipes, ate through concrete, and eventually cooled enough to solidify.

THE ELEPHANT'S FOOT

The spot where the corium solidified wouldn't be discovered until December in 1986. To contain the fallout, a large concrete enclosure named the sarcophagus was built on the site. Access points were left in the sarcophagus for researchers.

During one such research trip, their equipment registered levels of radiation so high that it would kill anyone who got too close for more than a few seconds. In order to see what was causing the readings, the scientist attached a camera to a wheeled contraption and rolled it in the direction the readings were emanating from. What they saw was dubbed the Elephant's Foot.

The Elephant's Foot is so deadly that spending only 30 seconds near it will result in dizziness and fatigue. Two minutes near it and your cells will begin to haemorrhage. By the time you hit the five-minute mark, you're a goner.

Even after 30 years, the foot is still melting through the concrete base of the power plant. Its existence makes the city uninhabitable to humans for at least the next 100 years. If it melts down into a source of ground water, it could trigger another explosion or contaminate the water of nearby villages.

And yet, in spite of the Elephant Foot's toxic presence in Chernobyl, something strange is happening.

ANIMALS FLOURISH

Biologists from the University of Georgia set up cameras in the Belarus evacuation zone to try and track animal activity in the area. What they found was surprising.

Many different kinds of animals aren't simply living in the irradiated area; they're thriving in it. The cameras spotted grey wolves, red foxes, wild boars, moose, and deer.

It's not that the area isn't still dangerous to humans, but instead, the animal life seems to have found a way to thrive in spite of it. And even more important than that, the flourishing animal life shows just how destructive the presence of human beings can be on the animal population of any given area.

Whatever the reason, the area of the Chernobyl disaster has become a kind of wildlife refuge for many different species of animals. At least some small amount of good was able to come from one of the worst disasters in the last three decades.

[AQA, 2018]

Section A style questions

1 **Source A** describes the Chernobyl accident. The Chernobyl accident occurred during tests on the Chernobyl 4 reactor in April 1986.

  1.1 State the **two** main reasons for the accident according to Source A. [2]

  1.2 Give **two** reasons why the information in Source A is likely to be more valid than information from a newspaper article. [2]

2 An exclusion zone was set up immediately following the accident. Thousands of local residents were relocated. In the years that followed, this exclusion zone was extended and more people relocated.

Use **Source A** to answer the following questions.

  2.1 Calculate the percentage increase in the size of the exclusion zone after it had been extended. [1]

  2.2 Extending the exclusion zone helped to protect people. Give **one** reason why the exclusion zone helped to protect people. [1]

  2.3 **Source A** implies that the extension of the exclusion zone might not have been necessary. Give **one** reason why. [1]

Section B style questions

7 **Figure 7.1** represents data for electricity generation in the UK.

3 Experts carried out several studies to assess the effects of the Chernobyl accident on the contaminated areas.

  3.1 One of the experts may have been a toxicologist. Suggest **one** role of a toxicologist in studies of the Chernobyl accident. [1]

  3.2 Why was it difficult to assess the effects of the accident on the health of the local people? Use information from **Source A**. [1]

  3.3 How did the experts assess the effects of the accident on the health of the local people? [1]

4 Some of the sources may have been peer reviewed.

Describe the process of peer review. [3]

5 The diagram in **Source A** shows the main environmental pathways of human radiation exposure.

Discuss how different groups of people affected by the Chernobyl accident were exposed to radiation and the consequences of exposure to radiation.

Use the diagram in **Source A** and evidence from **Sources A** and **B** in your answer.

The Quality of Written Communication will be assessed in your answer. [9]

6 Use Source B to answer the following questions:

  6.1 Suggest who the target audience for Source B might be. [1]

  6.2 Discuss how effectively Source B engages its target audience. [2]

  6.3 Give one reason why the information in Source B might not be valid. [1]

  6.4 Source B reports that biologists have found animals thriving in the exclusion zone around Chernobyl. Suggest why animals are thriving in the exclusion zone. [1]

  6.5 Suggest two roles of a biologist in the University of Georgia study referred to in Source B. [2]

Electricity generation in the UK between 2014 and 2016

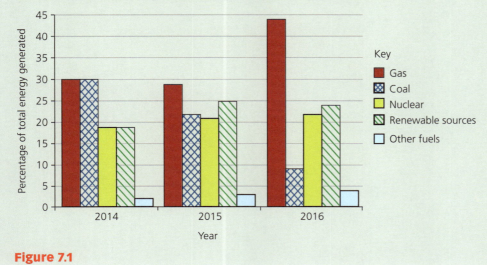

Figure 7.1

My Revision Notes: AQA Applied Science Suitable for Level 3 and Level 3 Extended Certificates

7.1 Suggest **two** reasons why scientists might use a chart such as **Figure 7.1** to represent data for electricity generation. [2]

7.2 Give **three** examples of 'renewable sources'. [3]

7.3 The use of energy from renewable sources in the UK was lower in 2016 than it was in 2015.

Suggest **three** reasons why the use of energy from renewable sources was lower in 2016. [3]

8 **Table 8.1** shows data for energy generated by nuclear power in different countries in 2016.

Use the data in **Table 8.1** to answer the following questions.

Table 8.1

Country	Energy from nuclear power / billion kWh	Percentage of total energy generated	Number of reactors	Approximate area of country / million km^2
USA	805.3	19.7	99	9.6
France	384.0	72.3	58	0.7
China	210.5	3.6	36	9.6
South Korea	154.3	30.3	25	0.1
Canada	97.4	15.6	19	10.0
Ukraine	81.0	52.3	15	0.6
UK	65.1	21.2	15	0.2

8.1 Which country relied more on nuclear power in 2016 than any other country? [1]

8.2 Give **one** piece of evidence from **Table 8.1** to support your answer to Question 8.1. [1]

8.3 The USA and China have approximately the same area in millions of km^2.

Compare the USA's and China's use of nuclear power to generate energy.

Use data from **Table 8.1**. [3]

8.4 Calculate the total electrical energy generated in the UK in 2016 to the nearest billion kWh. [2]

Check your understanding and progress at **www.hoddereducation.co.uk/myrevisionnotes**

Unit 4 The human body

The digestive system and diet

Humans rely on food for their energy. Most of the food we eat, however, contains chemicals too complex to enter the bloodstream. These complex organic chemicals must be broken down into simple inorganic chemicals to pass through the gut wall. This is the process of digestion and the human digestive system has a number of associated organs with specialised functions. For a healthy body we need a balanced diet with nutrients in the proper proportions. If the diet contains too much or too little of certain components, health problems can arise.

> **Making links**
>
> You can learn more about the way in which food is broken down to release energy by respiration in Unit 1, page 23.

The digestive system – structure and functions

REVISED ○

The structure of the human digestive system is shown in Figure 4.1, along with the functions of the different parts.

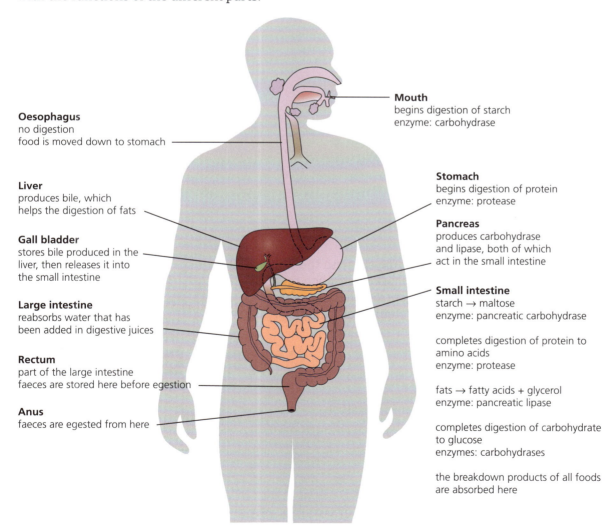

Oesophagus
no digestion
food is moved down to stomach

Liver
produces bile, which
helps the digestion of fats

Gall bladder
stores bile produced in the
liver, then releases it into
the small intestine

Large intestine
reabsorbs water that has
been added in digestive juices

Rectum
part of the large intestine
faeces are stored here before egestion

Anus
faeces are egested from here

Mouth
begins digestion of starch
enzyme: carbohydrase

Stomach
begins digestion of protein
enzyme: protease

Pancreas
produces carbohydrase
and lipase, both of which
act in the small intestine

Small intestine
starch → maltose
enzyme: pancreatic carbohydrase

completes digestion of protein to
amino acids
enzyme: protease

fats → fatty acids + glycerol
enzyme: pancreatic lipase

completes digestion of carbohydrate
to glucose
enzymes: carbohydrases

the breakdown products of all foods
are absorbed here

Figure 4.1 The structure of the human digestive system

The digestion of food involves two processes:

1 *Mechanical digestion* is the physical break up of food into smaller pieces, which increases the surface area for enzymes to work. It is carried out in the mouth (chewing) and by the muscular contractions of the gut known as peristalsis.

2 *Chemical digestion* is the chemical breakdown of the complex molecules in food to simpler substances. Enzymes act as catalysts for these chemical reactions.

Hydrolysis and condensation reactions

The complex organic molecules in food are broken down by digestion into simple inorganic substances which can be absorbed through the gut wall and into the bloodstream. These substances can later be reassembled into complex molecules again for use in the body. The molecules are broken down by *hydrolysis reactions* and built up by *condensation reactions*.

In a hydrolysis reaction, a water molecule is inserted, breaking a bond. The bonds concerned are:

+ *glycosidic* bonds in carbohydrates
+ *peptide* bonds in proteins
+ *ester* bonds in lipids.

The way these bonds are broken is shown in Figure 4.2.

Peptide bond:

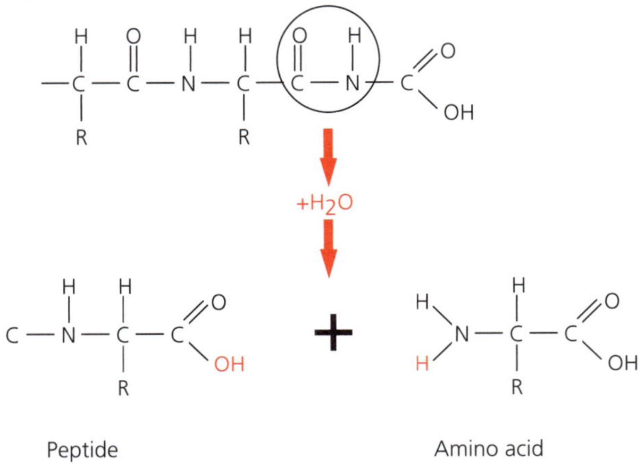

Hydrolysis of peptide bond

Figure 4.2 Hydrolysis reactions

A condensation reaction is the reverse of hydrolysis, with the components joining and water being eliminated as a result. Condensation reactions occur during assimilation, when simple sugars are built into complex carbohydrates, amino acids are built into proteins, and glycerol and fatty acids form lipids.

> **Now test yourself** TESTED ⚪
>
> 1 How is food moved along the gut?
> 2 Explain how mechanical digestion assists chemical digestion.
> 3 If amino acids are joined through condensation reactions, what will the products be?

Enzymes are essential to digestion

All the chemical reactions in digestion are catalysed by enzymes. Any given enzyme is specific, i.e. it will only work on a single substance or group of very similar substances. Different enzymes are therefore needed to digest carbohydrates, fats and proteins.

Check your understanding and progress at **www.hoddereducation.co.uk/myrevisionnotes**

Catalyst A substance which speeds up a chemical reaction and is left unchanged at the end of that reaction.

Enzyme A protein molecule which catalyses a chemical reaction in the body.

Peristalsis Rhythmic muscular contractions of the digestive tract (oesophagus, stomach and intestines). Peristalsis moves food through the gut and is involved in mechanical digestion.

Assimilation The processes by which an organism incorporates absorbed nutrients into the body.

Exam tip

Condensation is a term used in everyday language to describe the formation of water on cold surfaces. Use this to remember that water is formed in condensation reactions.

- *Carbohydrase* enzymes act in the mouth and small intestine to break down carbohydrates into simple sugars.
- *Protease* enzymes act in the stomach and small intestine to break down proteins into amino acids.
- *Lipase* enzymes act in the small intestine to break down lipids into fatty acids and glycerol.

Enzymes are not the only digestive fluids

Apart from enzymes, other digestive fluids act on the food as it passes down the gut. These fluids are described in Table 4.1.

Table 4.1 Digestive fluids

Fluid	Produced in	Function
Hydrochloric acid	Stomach	Kills bacteria. It also provides a suitable pH for pepsin enzyme, a protease produced by the stomach which has an optimum pH of 2.
Bile	Liver. Stored in, and released from, the gall bladder. Acts in the small intestine.	Emulsifies fats. Emulsification is the breakdown of large droplets into small ones. This provides an increased surface area for lipase enzyme to act on the fats.
Mucus	All parts of the gut	Lubricates the gut wall to allow food to pass down smoothly. Protects the wall of the gut from the action of protease enzymes and (in the stomach) hydrochloric acid.

Absorption in the small intestine   REVISED ○

The small intestine is divided into two sections, the *duodenum* (the first loop of the intestine after it leaves the stomach) and the *ileum*. Digestion is completed in the duodenum and the digested materials are absorbed in the ileum.

The structure of the small intestine is adapted to maximise absorption. The wall has numerous projections called *villi* (singular: villus), which greatly increase the surface area for absorption. The structure of a villus is shown in Figure 4.3.

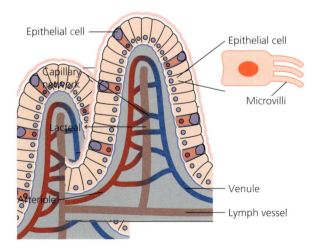

Epithelial cell
Epithelial cell
Capillary network
Microvilli
Lacteal
Venule
Arteriole
Lymph vessel

Figure 4.3 The structure of a villus

My Revision Notes: AQA Applied Science Suitable for Level 3 and Level 3 Extended Certificates

Structural adaptations of the villi are as follows:

+ The outer layer (epithelium) is only one cell thick, so there is a short diffusion pathway to the blood vessels.
+ The epithelial cells have microvilli on their surface, which further increase the surface area for absorption.
+ There is a good blood supply to the inside of each villus, to absorb glucose and amino acids.
+ *Lacteals* run into each villus. These are part of the body's lymphatic system, which eventually drains into the blood system. The lacteals absorb glycerol and fatty acids.

Absorption of glucose and amino acids takes place in the ileum

Glucose and amino acids are absorbed in the ileum by a process known as *co-transport*. Co-transport is a mechanism by which two substances are carried through the cell membrane by a single carrier (either in the same direction or in opposite directions). In this case, both glucose and amino acids travel (separately) through the membrane with sodium ions. The procedure for glucose is as follows:

1 Sodium ions (Na^+) are actively transported from the intestinal cell into the blood. This lowers the concentration of Na^+ in the cell.
2 Na^+ enters the cell from the gut down a diffusion gradient via a carrier, and this generates energy which allows the same carrier to actively transport glucose into the cell.
3 As the concentration of glucose in the cell rises, glucose is able to move into the blood, down a concentration gradient, by facilitated diffusion.
4 Na^+ also moves into the blood, but at this stage it is not co-transported with glucose.

Amino acids are absorbed in the same way as glucose, but different carriers are involved.

Gastrin is produced in the stomach

Gastrin is a hormone that stimulates growth of the lining of the stomach, movement of the stomach and secretion of hydrochloric acid (HCl) into the stomach. It is secreted by cells in the stomach wall and the duodenum, and travels in the bloodstream to the cells that produce HCl. Its secretion is stimulated by the presence of protein in the stomach, stretching of the stomach wall or a raised pH in the stomach.

Gastrin levels are sometimes monitored in patients. An increased gastrin level is thought to cause some stomach cancers, and some tumours in the stomach produce gastrin. Gastrin monitoring can therefore be used to assess risk and for cancer diagnosis.

> **Now test yourself** TESTED ◯
>
> 7 What is the function of the lacteals in the intestinal villi?
> 8 In terms of its role in the intestine, explain why low levels of sodium could lead to a feeling of tiredness.
> 9 The small intestine comes after the stomach in the gut, yet gastrin produced in the small intestine affects the stomach. Explain how this is possible.

Macro- and micronutrients are needed by the body

A variety of nutrients is required by the body. Some, known as *micronutrients*, are only needed in minute quantities and an excess of these may even be harmful. Those required in larger quantities are called *macronutrients*. Table 4.2 shows nutrients required, their functions and deficiency symptoms.

Epithelium The outermost layer of cells in animal tissues.

Lymphatic system Part of the immune system, which also maintains fluid balance and plays a role in absorbing fats and fat-soluble nutrients. It consists of a network of vessels that run through the whole body and drain into the blood system at the subclavian veins in the neck.

Microvilli Microscopic projections of a cell membrane.

Making links

You can learn more about active transport and facilitated diffusion in Unit 1, page 15.

Check your understanding and progress at **www.hoddereducation.co.uk/myrevisionnotes**

Table 4.2 Nutrients required by the body

Nutrient	Type	Function	Effects of deficiency	Sources
Carbohydrates	Macro	Main energy supply	Lack of energy; nausea; dizziness; constipation; bad breath; dehydration	Cereals, rice, pasta, bread, foods made with flour
Proteins	Macro	Growth and repair of tissues	Poor growth in children; hair, nail and skin problems; loss of muscle mass; brittle bones	Meat, fish, eggs, cheese, beans
Lipids	Macro	Energy store; insulation	Dryness of skin, hair and eyes; feeling cold; constant hunger; hormonal problems	Dairy foods, oily fish, red meat
Sodium	Macro	Helps to maintain body fluid concentration; needed for functioning of nerves and muscles	Lethargy; confusion; altered personality. Excess sodium is more common than deficiency.	Salt (many foods have added salt)
Calcium	Macro	Essential ingredient of bones and teeth; needed for nerve function	Muscle problems; dental problems; brittle bones	Dairy foods; green leafy vegetables
Iron	Micro	Necessary for the manufacture of red blood cells	Anaemia	Meat; liver; wholegrains; beans
Vitamin C	Micro	Necessary for growth, development and repair of body tissues	Scurvy (easy bruising, tiredness, muscle and joint pains)	Citrus fruits; peppers; strawberries; broccoli; Brussels sprouts; potatoes
Vitamin D	Micro	Plays a role in the nervous, muscle and immune systems; needed for proper absorption of calcium	Rickets (soft bones in children); brittle bones; weakened immune system	Oily fish; red meat; liver; egg yolks; made in the skin in response to sunlight

Now test yourself   TESTED ◯

10 Suggest why pregnant women are sometimes advised to take iron supplements.

11 Bones are initially made from cartilage, which is softer and more flexible than bone. The cartilage is then 'calcified' by calcium to form the bone. Suggest why rickets tends to be a disease of children rather than adults.

Making links

You can learn more about salt deficiency and excess salt in Unit 1, in the section on Homeostasis, page 19.

The musculoskeletal system and movement

To live on land, organisms need some means of support, otherwise their bodies would collapse under their own weight. In humans and other vertebrates, this support is provided by the skeleton, made of bone. To move, however, two further features are needed: joints within the skeleton (as the bones are rigid) and muscles to move those joints. The bones with their joints and attached muscles make up the musculoskeletal system.

Structure and functions of the skeleton   REVISED ◯

The skeleton has two sets of bones

The skeleton is made up of two sets of bones, called the axial and appendicular skeletons.

✚ The axial skeleton is the skeleton's central axis, consisting of the skull, vertebral column and rib cage.
✚ The appendicular skeleton is the limbs, together with the bones of the shoulders and the pelvis.

The skeleton consists of two tissue types: bone and cartilage. In the embryo, the skeleton is initially laid down in cartilage which then hardens into bone

133

with the addition of calcium. Cartilage remains at the ends of many bones where they form joints, as it is softer than bone and creates less friction where bones come together.

The skeleton provides more than just support

The skeleton has the following functions:
+ *Support* – it forms a rigid framework for the body.
+ *Protection* – the skull, rib cage and pelvis protect internal organs.
+ *Movement* – the joints in the skeleton allow for movement, created by attached muscles.
+ *Marrow/blood cell production* – the soft bone marrow inside hollow bones is the main site for the manufacture of red blood cells.

Bone formation and resorption

REVISED

When an embryo is about 6 or 7 weeks old, the process of bone development (ossification) begins. There are two types of ossification: *intramembranous* and *endochondral*. Intramembranous ossification forms bone directly from connective tissue and endochondral ossification forms bone from pre-existing cartilage. Endochondral ossification occurs during development, but also when bones grow and when broken bones are repaired. New bone is laid down by specialised cells called *osteoblasts*.

In bones there are also cells called *osteoclasts*, which break down (or 'resorb') bone tissue and so release calcium. The usual reason for this is to maintain the blood concentration of calcium.

Bone is a living and dynamic tissue and can respond to increased stress by adding more bone tissue to make the bone stronger or remove it if the stress is lessened.

> **Ossification** The process of bone formation.
>
> **Resorption** The absorption into the circulation of material from cells or tissue.

Synovial joints allow movement between bones

As bones are rigid, there must be joints between them to allow movement. A *synovial joint* is the type of joint found between bones that move against each other (e.g. hip, shoulder, elbow and knee). The general structure of a synovial joint is shown in Figure 4.4. The cartilage is softer than bone so the bones do not grate together, and the synovial fluid contained within the capsule lubricates the joint. Synovial joints come in several types:
+ *Gliding*. The bones slide past one another, e.g. some joints in the wrists and ankles.
+ *Hinge*. The joint moves like a hinge, in one plane only, e.g. elbow and knee.
+ *Ball and socket*. A 'ball shaped' end of one bone fits into a cup-like depression in another bone. The ball and socket can move in three planes (more than any other type of synovial joint).
+ *Pivot*. A rounded end of one bone fits into a ring formed by another bone, e.g. between the vertebrae in the neck, and in the wrist, allowing the hand to be twisted.

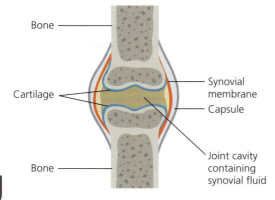

Figure 4.4 Synovial joint

Check your understanding and progress at **www.hoddereducation.co.uk/myrevisionnotes**

The structure of muscles allows them to contract

Skeletal muscles have a complex structure, shown in Figure 4.5. The smallest unit of muscle is the myofibril, which is the contractile element. The structure of a myofibril is shown in Figure 4.6.

Myofibril A contractile fibre in skeletal muscle.

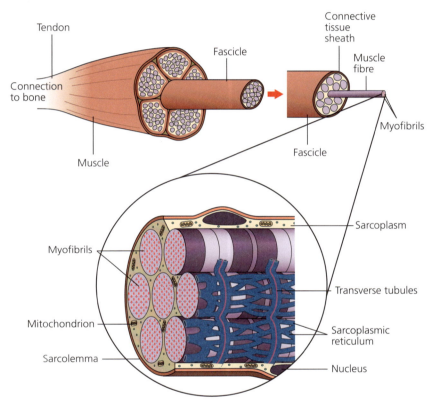

Figure 4.5 Structure of skeletal muscle

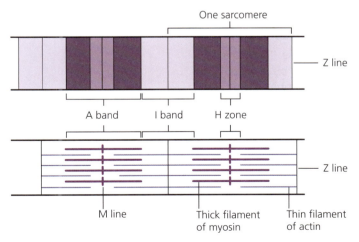

Figure 4.6 Structure of a myofibril

Exam tip

The structure of the muscle shown in Figure 4.5 is for context only. In the exam, you will only be tested on the structure of a myofibril.

135

Myofibrils are made up of two types of myofilament: thin filaments made of the protein *actin* and thick filaments made of the protein *myosin*. The thin and thick filaments partially overlap. The myofibril has repeating units called sarcomeres. The arrangement of the myofilaments gives the appearance of dark and light bands. The different regions of the myofibril are shown in Figure 4.6. The Z line is a protein disk that anchors the actin filaments.

> **Myofilament** Protein filament found in a myofibril. There are two types: actin and myosin.
>
> **Sarcomere** The structural unit of skeletal muscle.

The sliding filament theory of muscle contraction

The sliding filament theory of muscle contraction proposes that muscle contraction is brought about by the thick and thin myofilaments sliding past one another, shortening the sarcomere. To understand this mechanism, we need to know more about the structure of the thick and thin filaments, as shown in Figure 4.7.

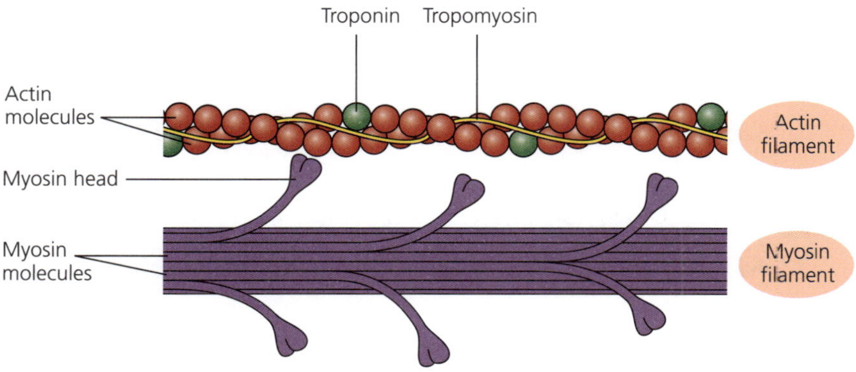

Figure 4.7 Structure of the thin and thick filaments

Summary of the sliding filament mechanism

+ Myosin filaments consist of a 'head' and a 'tail'.
+ The heads can attach to actin and move it.
+ The heads can also bind ATP.
+ Actin has binding sites for myosin.
+ When relaxed, the actin–myosin binding sites are blocked by a protein called *tropomyosin*, which is held in place by another protein, *troponin*.
+ During contraction, the troponin shifts and moves the tropomyosin, exposing the actin–myosin binding sites.
+ The myosin heads bond to the actin and form cross-bridges. The myosin heads flex and pull the actin filament along the myosin filament. The cross-bridges form and break up to 100 times each second.
+ The myosin heads detach when fresh supplies of ATP arrive, and the process is repeated for as long as the stimulus for contraction continues.

Calcium is important during muscle contraction

Calcium ions play an important part in the contraction process. They are released into the muscle when a nerve stimulus arrives. They bind with troponin and this causes the tropomyosin to change shape, unblocking the myosin binding sites on the actin filaments so that cross-bridges can form. When nerve impulses cease, the calcium ions are actively transported out of the sarcoplasm (the name given to the cytoplasm of the muscle cells) and the muscle then relaxes.

Check your understanding and progress at **www.hoddereducation.co.uk/myrevisionnotes**

The sliding filament theory is summarised in Figure 4.8.

(a)
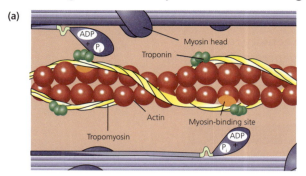
The resting state. The muscle is relaxed.

(b)
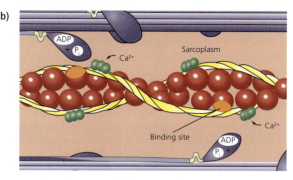
An impulse arrives and calcium ions flood into the myofibril. These ions bind to the troponin, which moves the tropomyosin out of the myosin-binding sites.

(c)
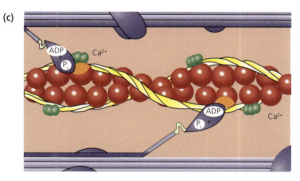
The myosin heads attach to the exposed binding sites.

(d)
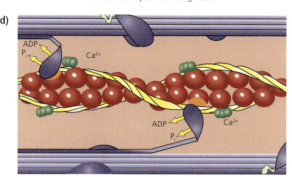
The release of ADP and inorganic phosphate causes the heads to move and pull the actin along.

(e)
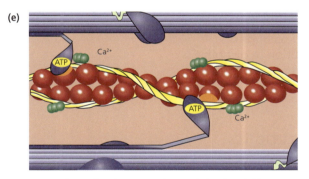
Fresh supplies of ATP enter the myosin heads and this breaks the connection with the binding sites.

(f)
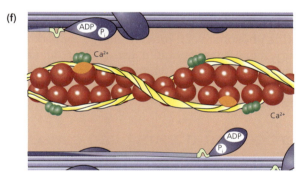
The hydrolysis of ATP to ADP and P_i returns the myosin heads to their starting positions. If calcium ions are still present, each myosin head will then immediately bind to the next myosin-binding site.

Figure 4.8 The stages of contraction in a myofibril. P_i represents inorganic phosphate

Now test yourself

TESTED ◯

16 Name the two types of muscle myofilament.
17 What is the role of ATP in the sliding filament theory of muscle contraction?
18 How does troponin cause the binding of actin to myosin?

There are two types of muscle fibre, called slow-twitch and fast-twitch. Fast-twitch fibres are 'white' (i.e. pale) and slow-twitch fibres are red. They have different properties which suit them to their function.

✦ Slow-twitch fibres are good for prolonged but low-intensity activity, e.g. maintenance of body posture, movement of food through the gut. They are slower to contract than fast-twitch fibres, but are resistant to fatigue so they can contract consistently over a long period of time.

✦ Slow-twitch fibres produce ATP at a slower rate than fast-twitch fibres and so cannot contract as powerfully. Fast-twitch fibres produce ATP rapidly and so contract much more quickly and powerfully.

+ Slow-twitch fibres have more resources than fast-twitch fibres, as they store glycogen and have the ability to respire fat stores. Fats release energy slowly and so would not be of value in fast-twitch fibres.
+ Fast-twitch fibres fatigue more quickly, because intense activity requires some anaerobic respiration in the muscle fibres, which generates lactate.

The differences between the two types of fibre are shown in Table 4.3.

Table 4.3 Differences between fast-twitch and slow-twitch muscle fibres

	Slow-twitch fibres	Fast-twitch fibres
Suitable for	Endurance	Strength and power
Type of respiration	Aerobic	Aerobic and anaerobic
Blood supply	Many blood vessels	Fewer blood vessels
Density of mitochondria	High	Low
Amount of myoglobin	Low	High
Contraction rate	Slow	Fast

Myoglobin is a pigment in muscles that holds oxygen and releases it only when oxygen levels are extremely low. It effectively acts as a sort of emergency store of oxygen which will be released during strenuous activity.

Creatine phosphate is an ATP 'store'

The ATP supplies in muscles can only provide energy for 1 or 2 seconds of intense activity. To prolong this, both fast-twitch and slow-twitch fibres contain an ATP 'store' in the form of *creatine phosphate*. The breakdown of creatine phosphate provides two resources:
+ The breakdown releases energy which can be used to make ATP from ADP.
+ Phosphate is released which is also needed in the manufacture of ATP.

Creatine phosphate can provide energy for about a further 10 seconds of activity. It can be regenerated, using ATP from aerobic respiration to combine phosphate with creatine when the period of activity ends. Creatine is formed during the breakdown of creatine phosphate and is also found in cytoplasm and intercellular fluid.

Exercise and training can cause muscle type to change

The proportions of fast-twitch and slow-twitch muscles vary slightly in different people but these proportions can be changed by exercise.
+ Exercises involving a lot of power (e.g. sprinting, lifting heavy weights, bench presses and squats) will increase the proportion of fast-twitch muscle fibres.
+ Endurance training (e.g. long-distance running and cycling) will increase the proportion of slow-twitch fibres.

Athletes in training also use diet to maximise performance. Their diet is high in carbohydrates (wholegrain, rather than processed) for energy and protein (to build muscle), and they may also take dietary supplements. A common supplement is creatine, as this increases the store of creatine phosphate in muscles to provide extra energy.

Making links

You can learn more about ATP and cellular respiration in Unit 1, page 23.

Now test yourself TESTED

19 Muscles attached to the spine help to maintain posture. Which type of muscle fibre (fast or slow twitch) is likely to be more common in these muscles? Explain your answer.
20 Suggest a reason why slow-twitch muscles do not contain much myoglobin.
21 When training, sprinters tend to use heavy weights over short periods of time, whereas long-distance runners use lighter weights over a longer period. Suggest the reasons for this difference.

Check your understanding and progress at **www.hoddereducation.co.uk/myrevisionnotes**

Oxygen transport and physiological measurements

Haemoglobin and oxygen transport

REVISED ○

Water-soluble substances can easily be transported around the body dissolved in blood plasma, but oxygen does not dissolve very well in water and so it needs to be transported in a different way, attached to *haemoglobin* (Hb) in red blood cells to form *oxyhaemoglobin*.

Haemoglobin (Figure 4.9) is a protein which consists of four polypeptide chains (referred to as either α chains or β chains – two of each) each containing a *haem group*. A haem group is a chemical group which contains an iron ion at its centre.

> **Polypeptide** A chain of amino acids of insufficient length to be called a protein.

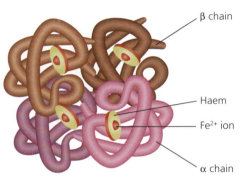

β chain

Haem

Fe^{2+} ion

α chain

Figure 4.9 The haemoglobin molecule

The ability of haemoglobin to bind with oxygen (called its 'affinity' for oxygen) varies according to the partial pressure of oxygen in its surroundings. The relationship is shown by an *oxygen dissociation curve* (Figure 4.10).

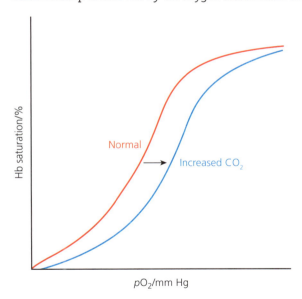

Hb saturation/%

Normal

Increased CO_2

pO_2/mm Hg

Figure 4.10 Oxygen dissociation curve

The normal curve shows why haemoglobin can absorb oxygen in the lungs but give it up again in the tissues. The lungs have a high partial pressure of oxygen (pO_2). This allows the haemoglobin to become highly saturated (i.e. absorb a lot of oxygen). When the red blood cells reach the tissues, these tissues have a much lower pO_2 because they are respiring. At these lower values, haemoglobin can only achieve a low % saturation, and so must release oxygen.

Notice that an increase in carbon dioxide concentration shifts the curve to the right (this is known as the *Bohr effect*) so that, at any level of oxygen saturation,

139

the haemoglobin's affinity for oxygen decreases. Actively respiring tissues (which need oxygen) will be producing carbon dioxide, and this will mean that even more oxygen is released to the tissues.

Oxygen transport and high-altitude training

Endurance athletes (e.g. long-distance runners) must maintain aerobic respiration throughout their event. They can be assisted by high-altitude training. At high altitude, the partial pressure of oxygen decreases and the body adapts to this by producing more red blood cells. This extra haemoglobin is maintained for about 10–14 days after returning to lower altitude, which helps athletes' stamina in endurance events.

There is some evidence that high-altitude training may also improve the efficiency of muscles in using oxygen.

> **Making links**
>
> You can learn more about aerobic and anaerobic respiration in Unit 1, page 23.

Oxygen saturation

The pulse oximeter measures oxygen saturation

Oxygen saturation can be measured using a *pulse oximeter*, a small device which attaches to the fingertip, ear lobe or toe. It sends out electromagnetic radiation of two wavelengths (red and infrared) and detects the difference in absorption of these two wavelengths, which is affected by how much oxygen there is in the blood. The percentage concentration of oxygen in the blood is displayed on a digital read-out. The normal range of oxygen saturation ($SaO_2\%$) in a healthy person is between 95 and 99%. The $SaO_2\%$ can be used to monitor or assess patients with lung conditions (e.g. emphysema, cystic fibrosis) and is routinely used to monitor patients in critical care units, in case help with breathing is required.

Emphysema: the lungs become less elastic

In emphysema, the walls of the alveoli (air sacs) in the lungs break down and the elasticity of the lungs is reduced. This means that less oxygen is absorbed into the blood at each breath. Measurement of the $SaO_2\%$ can be used to assess the severity of the damage.

Cystic fibrosis: thicker mucus in the lungs

Patients with the lung condition cystic fibrosis have thicker mucus in the lungs so oxygen cannot be absorbed easily. This leads to a low concentration of oxygen in the blood, which causes tiredness and weakness and can lead to lung damage. It is recommended that $SaO_2\%$ is maintained above 92% and so regular monitoring is necessary.

Blood pressure and its measurement

Blood pressure is the force that your heart uses to move blood around your body. It is measured in millimetres of mercury (mmHg) and is given as two figures, *systolic* pressure (the pressure of the heartbeat) and *diastolic* pressure (the pressure when the heart is relaxed between beats). Ideal blood pressure is between 90/60 mmHg and 120/80 mmHg, the first figure being the systolic pressure, the second the diastolic.

Blood pressure is measured using a *sphygmomanometer* composed of an inflatable cuff to collapse and then release the artery under the cuff, and a mercury or aneroid manometer which measures the pressure. The cuff is inflated to block off the artery (giving a systolic reading) and then deflated again to give the diastolic reading. Mercury manometers are the most accurate but because of the dangers of mercury they are only used in clinical situations. Automated electronic sphygmomanometers are available for home use.

High blood pressure (*hypertension*) puts a strain on artery walls and increases the risk of heart attack and strokes.

Low blood pressure (*hypotension*) can lead to dizziness and fainting, nausea and fatigue.

The structure and function of the nervous system and brain

The nervous system   REVISED ◯

The nervous system co-ordinates the body by sending electrical impulses via nerves. The nervous system consists of two parts, the *central nervous system* (CNS) and the *peripheral nervous system* (PNS). The central nervous system is the brain and the spinal cord, while the peripheral nervous system is made up of all the nerves leading from the CNS.

The nervous system is also divided into two functional parts: the *somatic* nervous system and the *autonomic* nervous system. The somatic nervous system controls voluntary activity and the autonomic nervous system controls involuntary functions (e.g. heart rate, movement of the gut). The autonomic nervous system is split into the *sympathetic* and *parasympathetic* nervous systems, which generally have opposite effects. The sympathetic nervous system has a role in responding to threatening situations, increasing the heart rate and dilating the pupils. The parasympathetic system decreases heart rate and stimulates movement of the gut.

The brain   REVISED ◯

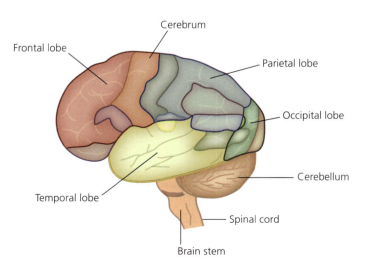

Figure 4.11 The brain

The brain is divided into three main areas: the *cerebral cortex*, the *cerebellum* and the *brain stem*. The cerebral cortex has four lobes, each with different functions.

+ The *frontal lobe* is associated with reasoning, planning, movement, emotions and problem solving.
+ The *parietal lobe* is associated with movement, orientation and recognition.
+ The *occipital lobe* is associated with visual processing.
+ The *temporal lobe* is associated with perception and recognition of auditory stimuli (hearing), memory and speech.

The main role of the cerebellum is in controlling skeletal muscle for fine movement, the co-ordination of movement, posture and balance. It does not initiate movement but co-ordinates it in response to information from a variety of sense organs.

The brain stem is responsible for controlling and maintaining vital (involuntary) functions such as breathing, blood pressure and heart rate.

Brain function indicates where the brain has been injured

As each part of the brain has a specific set of functions, it is possible to identify the area which has been damaged by the symptoms associated with the damage.

Damage to the cerebral cortex will result in faults in sensory or voluntary activities. For example, if the occipital lobe is damaged, vision may be affected, or the patient may be unable to recognise what they see. Damage to the cerebellum will affect balance and the co-ordination of movement. Damage to the brain stem can result in breathing difficulties or difficulty in swallowing, for example.

> **Now test yourself** TESTED ⬤
>
> 28 Name the four lobes of the cerebral cortex.
>
> 29 The vagus nerve connects the brain stem to the heart and is part of the parasympathetic system.
> > a Is the vagus nerve part of the central nervous system or the peripheral nervous system?
> > b What effect will impulses from the vagus nerve have on the heart?
>
> 30 Which part of the brain will be active when a person is solving a maths problem?
>
> 31 A patient with brain damage cannot find their way back home if they leave the house. Which part(s) of the brain are most likely damaged? Give reasons for your answer.

> **Exam tip**
>
> The brain is complicated and damage to a specific area can involve many different symptoms which cannot all be listed here. In an exam, you are likely to be given a symptom and asked to identify which area it relates to, from your knowledge of the general functions of each area.

Nerve impulses

Nerve structure   REVISED ⬤

Nerves are composed of many neurones (nerve cells). The neurones are of two types: *motor neurones* take impulses from the central nervous system out to effectors (muscles or glands); *sensory neurones* carry impulses from sense organs to the central nervous system. These two types of neurone have different structures, as shown in Figure 4.12.

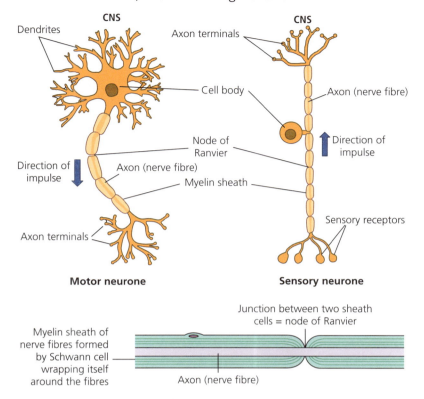

Figure 4.12 The structure of a motor neurone and a sensory neurone

In the body there are three types of nerve: *sensory nerves*, containing only sensory neurones; *motor nerves*, containing only motor neurones; and *mixed nerves*, containing both types of neurone.

Resting potential – when there is no impulse in the nerve

A nerve impulse is an electrical current travelling down a neurone. Neurones generate an impulse by pumping positively charged sodium and potassium ions across their membrane. When there is an unequal distribution of ions on either side of the membrane, the charge difference is called a *membrane potential*. This can be a resting potential (when there is no impulse) or an action potential when an impulse is generated. At resting potential, the inside of the neurone is more negative than the outside. The resting potential is approximately –70 mV. The maintenance of a resting potential is created by sodium–potassium pumps in the neurone membrane.

The sodium–potassium pump is a carrier protein in the membrane which exchanges sodium and potassium ions by active transport, using ATP. It expels three Na^+ ions and at the same time takes in two K^+ ions. Some K^+ ions will leak back out down a concentration gradient, but in the resting state no Na^+ ions can get back in. This causes an electrochemical gradient where the interior of the cell is more negative than the outside, because there are more positively charged ions in the extracellular environment and more negatively charged ions inside the cell.

Action potential occurs when a stimulus arrives at a neurone

When a stimulus arrives at a neurone, an action potential results. This is a reversal of the resting potential and results in a current (an impulse) travelling along the neurone. The action potential results in the following way.

Exam tip

Make sure you do not confuse axon terminals (which are connected to the axon) with dendrites (which are attached to the cell body).

1. When a stimulus arrives, sodium channels in the membrane open, allowing Na$^+$ ions to enter the neurone down a concentration gradient.
2. The increase in positive ions inside the cell *depolarises* the membrane potential (i.e. decreases the difference between the inside and the outside).
3. If the depolarisation reaches the threshold potential (around −50 to −55 mV), extra voltage-gated sodium channels open, causing a rush of Na$^+$ ions into the neurone.

Electrochemical gradient A situation where there is a gradient of electrical charge across a membrane.

Threshold potential The membrane potential which, when reached, initiates a nerve impulse.

Voltage-gated channel A channel protein in the cell membrane which is only open at a limited range of membrane potentials.

Depolarisation

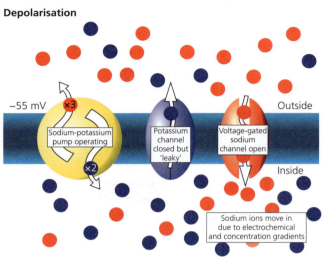

Beginning of repolarisation

End of hyperpolarisation

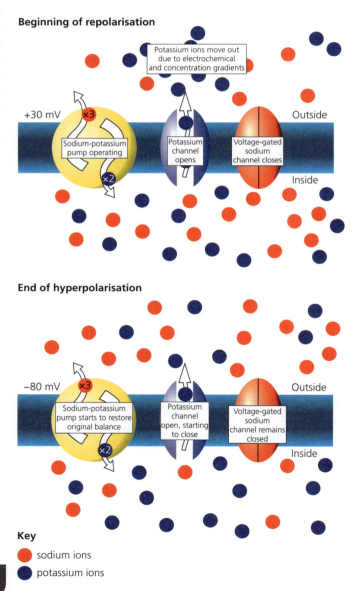

Key

🔴 sodium ions

🔵 potassium ions

Figure 4.13 Events during depolarisation, repolarisation and hyperpolarisation

Check your understanding and progress at **www.hoddereducation.co.uk/myrevisionnotes**

4 The potential across the membrane rapidly reverses to become more positive inside than outside.

5 Eventually, this change in potential causes extra (voltage-gated) potassium channels to open, allowing K⁺ ions to leave the cell. This causes the membrane potential to decrease once more.

6 As a result of this, the voltage-gated sodium channels start to close.

7 The membrane repolarises and then drops slightly below the resting membrane voltage. This occurs because there are more potassium channels open than at the resting state.

8 At about –90 mV, the 'extra' potassium channels close and the sodium-potassium pump restores the resting potential.

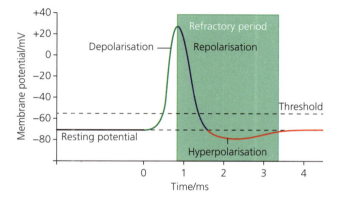

Figure 4.14 Graph of membrane potential against time during the production of an action potential

The importance of the myelin sheath

REVISED ○

An action potential in part of a neurone acts as a stimulus for the adjacent membrane to depolarise. Where a myelin sheath is present, the impulse does not pass straight along the axon, but moves in a series of jumps. The myelin sheath acts as electrical insulation but it has a series of gaps in it, called *nodes of Ranvier*. The impulse 'leaks out' of these nodes and jumps along the surface of the myelin sheath to the next node. Moving this way, the impulse travels even more quickly than it would if it spread down the axon. Not all neurones have a myelin sheath. Where the axon is short, the speed increase would be minimal so these axons are unmyelinated, with the impulse travelling directly through the axon.

Now test yourself TESTED ○

32 In which direction (towards or away from the central nervous system) do motor neurones carry impulses?

33 When a stimulus arrives and the sodium channels of the neurone membrane open, why do sodium ions enter the cell rather than leave it?

34 The movement of which ion is responsible for repolarisation?

Synapses

REVISED ○

Nerve impulses need to pass from one neurone to another. This happens at junctions called synapses. At each synapse there is a small gap (the synaptic cleft); the impulse is carried across this gap by chemicals called neurotransmitters. The mechanism also involves calcium ions (Ca^{2+}) and is summarised in Figure 4.15.

Once an action potential has started in the postsynaptic neurone, the transmitter must be removed from the cleft; otherwise it will continue to stimulate the postsynaptic neurone. An enzyme breaks down the neurotransmitter and the breakdown products are absorbed back into the presynaptic knob for re-use. The enzyme used will depend on the neurotransmitter. The nervous system uses a range of neurotransmitters, including *acetylcholine*, *dopamine* and *serotonin*.

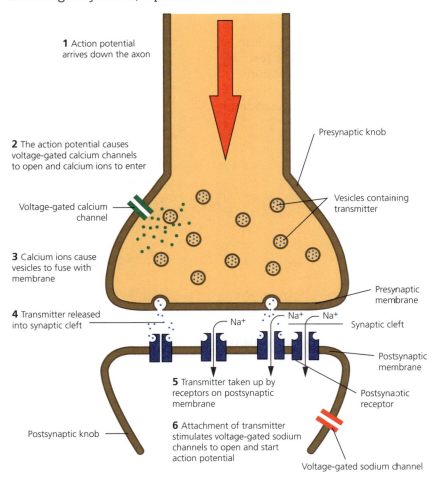

1 Action potential arrives down the axon

2 The action potential causes voltage-gated calcium channels to open and calcium ions to enter

Voltage-gated calcium channel

3 Calcium ions cause vesicles to fuse with membrane

4 Transmitter released into synaptic cleft

Presynaptic knob

Vesicles containing transmitter

Presynaptic membrane

Na+   Na+   Na+

Synaptic cleft

Postsynaptic membrane

5 Transmitter taken up by receptors on postsynaptic membrane

Postsynaptic receptor

Postsynaptic knob

6 Attachment of transmitter stimulates voltage-gated sodium channels to open and start action potential

Voltage-gated sodium channel

Figure 4.15 The events at a synapse

Synapses and disorders

REVISED

Problems with neurotransmitters can lead to certain disorders.

Alzheimer's disease is associated with a fall in the level of acetylcholine, which interferes with the action of synapses. This results in mental incapacity (loss of memory, inability to concentrate, speech difficulties, etc.). It is thought that bundles of tissue called plaques, which form in the brain in this disease, stimulate the action of the enzyme acetylcholinesterase, which catalyses the breakdown of acetylcholine. Treatments using inhibitors of acetylcholinesterase have had some success in slowing the development of Alzheimer's disease.

Parkinson's disease is a disease which causes the patient to gradually lose muscular control and co-ordination. It is linked with a fall in dopamine, a neurotransmitter in the brain. As the disease progresses and dopamine levels decrease, the symptoms of the disease become more apparent and the patient develops tremors, difficulty walking and other issues with movement. Levodopa, the most effective Parkinson's disease medication, is a chemical that is converted to dopamine in the brain.

Depression is linked with (but probably not caused by) low levels of the neurotransmitter serotonin. Serotonin is known to be associated with the feeling of pleasure. Selective serotonin reuptake inhibitors (SSRIs) are a common antidepressant treatment. They prevent serotonin from being reabsorbed at synapses. This means that serotonin levels stay high in the brain and this eases the feeling of depression. Symptoms of depression vary between individuals but generally involve feelings of unhappiness and hopelessness.

Check your understanding and progress at **www.hoddereducation.co.uk/myrevisionnotes**

Now test yourself

TESTED ○

35 What effect does a neurotransmitter have on the postsynaptic membrane in a synapse?

36 Why might low levels of calcium affect synapses?

37 Explain how drugs containing an inhibitor of acetylcholinesterase enzyme might benefit patients with Alzheimer's disease.

Summary

+ The digestive system consists of an association of organs, each of which has a specific role in chemical and/or mechanical digestion.
+ Carbohydrates, proteins and fats are broken down by hydrolysis reactions in digestion and may be reassembled into new substances by condensation reactions during assimilation.
+ Carbohydrases, proteases and lipases catalyse the breakdown of carbohydrates, proteins and lipids, respectively.
+ Hydrochloric acid in the stomach kills bacteria and provides the optimum pH for pepsin enzyme.
+ The release of hydrochloric acid in the stomach is stimulated by the hormone gastrin.
+ Bile from the liver emulsifies fats.
+ Mucus lines the digestive system and provides lubrication for the food moving through it.
+ The small intestine absorbs the products of digestion and its surface area for absorption is increased by the presence of villi.
+ Glucose and amino acids are absorbed into the bloodstream by the process of co-transport.
+ The body requires certain macronutrients and micronutrients for specific functions. These include carbohydrates, proteins, lipids, sodium, calcium, iron, vitamin C and vitamin D. Deficiencies of these macronutrients and micronutrients lead to specific diseases and disorders.
+ A balanced diet contains foods that can help to maintain healthy levels of macronutrients and micronutrients.
+ The skeleton is made up from cartilage and bone and is divided into the axial and appendicular skeleton.
+ The functions of the skeleton include support, protection, movement and the manufacture of blood cells by the bone marrow.
+ Bone is constantly remodelled by the processes of resorption and ossification.
+ The structure of synovial joints includes adaptations for movement.
+ Different types of synovial joint have different ranges of movement. These joints include gliding, hinge, ball and socket, and pivot joints.
+ The myofibrils of muscle include actin and myosin filaments.
+ The structure of myofibrils includes Z lines, A bands, H bands and I bands.
+ The sliding filament theory of muscle contraction describes how myosin and actin interact to cause contraction which requires energy from ATP.
+ Calcium ions play an important role in the muscle contraction mechanism.
+ Muscle fibres are either fast-twitch (white) or slow-twitch (red).
+ Fast-twitch fibres generate ATP very quickly and are used for short bursts of explosive action, while slow-twitch fibres produce ATP more slowly and are not very powerful.
+ Fast-twitch and slow-twitch fibres have adaptations which suit their functions.
+ The breakdown of creatine phosphate in muscles transfers energy and releases phosphate ions which are used to make ATP. The creatine phosphate is regenerated during aerobic respiration.
+ Exercise affects the proportion of fast-twitch and slow-twitch muscle fibres.
+ Athletes adapt their diet during training and may include the use of creatine supplements.
+ Oxygen does not dissolve well in plasma and so most oxygen is carried by haemoglobin in the red blood cells.
+ Haemoglobin is a protein with a tertiary structure.
+ The degree of oxygenation of haemoglobin depends on the partial pressure of oxygen.
+ The oxygenation of haemoglobin can be shown by an oxygen dissociation curve.
+ The presence of carbon dioxide assists in the dissociation of oxygen (the Bohr effect).
+ Training at high altitude affects oxygen transportation.
+ Pulse oximeters measure oxygen saturation. The normal range is between 95 and 99% and is expressed using SaO_2%.
+ Diseases such as emphysema and cystic fibrosis affect oxygen saturation levels.
+ A sphygmomanometer is used to measure blood pressure. High or low blood pressures affect health.
+ The nervous system is organised into the central nervous system and the peripheral nervous system.
+ The peripheral nervous system is organised into the somatic and autonomic nervous systems, which have different roles.
+ The autonomic nervous system is divided into the sympathetic and parasympathetic nervous systems, which tend to have opposite effects.
+ The brain is organised into the lobes of the cerebral cortex, the cerebellum and the brain stem.
+ The four lobes of the cerebral cortex each have specific roles.
+ The cerebellum controls skeletal muscle for fine movement, co-ordination, posture and balance.
+ The brain stem maintains vital functions such as breathing and heart rate.
+ Brain damage may result in symptoms that indicate the area of the brain that has been affected.
+ Sensory and motor neurones have different structures.

+ The movement of ions into and out of neurones creates the resting potential and the action potential. The sodium–potassium pump has a particular role in maintaining the resting potential.
+ The nerve impulse travels rapidly along a nerve fibre, and the structure of myelinated nerve fibres enables them to conduct impulses more quickly.
+ Impulses travel from one neurone to another across gaps called synapses, which have a structure which enables this process.

+ The transmission of an impulse across a synapse involves calcium ions and enzymes.
+ The body has a wide range of neurotransmitters, including acetylcholine, dopamine and serotonin.
+ Disorders arising from problems with neurotransmitters and synaptic transmission include Alzheimer's disease, Parkinson's disease and depression. Each of these disorders has specific symptoms.
+ Drugs used to treat these disorders affect synaptic transmission.

Exam practice

1 Rickets is a disease of children that is caused by a lack of vitamin D. The children have soft bones because cartilage in their bones is not sufficiently hardened with calcium. In the UK, rickets is more common in black and south Asian populations than in the white population. The incidence is particularly high in recent immigrants who maintain the diet they ate in their country of origin. Vitamin D is made in the skin in sunlight, and foods high in vitamin D include red meat, liver, oily fish and egg yolks.

   1.1 The table below shows some data from various sources about rickets in the UK population. Fill in the missing figures. [2]

Ethnic group	Number in the UK population	% of UK population	Number of cases of rickets	% of ethnic group with rickets
White	48 209 395	86	193	0.0004
South Asian	2 984 670	4.5		0.038
Black	1 864 890	3.3	1772	

   1.2 Explain why a shortage of vitamin D can cause soft bones. [2]

   1.3 The incidence of rickets in children living in the UK who have come from India, Pakistan and Bangladesh is higher than that of children still living in those countries and eating a similar diet. Suggest an explanation for this. [3]

   1.4 Vitamin D is a micronutrient. Name **one** other micronutrient needed in the diet. [1]

   1.5 The bulk of the human diet is made up of carbohydrates, proteins and lipids. State **one** use of each of these nutrients in the body. [3]

2 Muscles contain two types of fibre, fast-twitch and slow-twitch. The proportions of these fibres vary slightly among individuals at birth but can be modified by activities. The proportions can vary significantly in athletes training for different types of athletic event, as shown in the table below.

	Mean % fast-twitch fibres	Mean % slow-twitch fibres
Untrained person	50	50
Sprinter	80	20
Middle-distance runner	30	70
Long-distance runner	20	80
Weightlifter	60	40
Footballer	50	50
Cyclist	60	40

   2.1 What is the likely cause of the differences among individuals at birth? [1]

   2.2 The differences seen in the table are due to specialised training regimes. Suggest reasons why a sprinter, a long-distance runner and a cyclist need different proportions of the fibre types. [5]

   2.3 A footballer has roughly the same fibre proportions as an untrained person yet is highly trained. Suggest what differences would be noticed between the muscles of a footballer and an untrained person. [1]

Check your understanding and progress at **www.hoddereducation.co.uk/myrevisionnotes**

2.4 Suggest which of the following training exercises would be most suitable for a long-distance runner. [1]
   A Lifting small weights, with many repeats
   B Lifting heavy weights, with few repeats
   C Lifting heavy weights, with many repeats
   D Lifting small weights, with few repeats

2.5 Elite athletes sometimes take creatine supplements. Explain the benefits of this. [3]

2.6 A common injury in professional football is torn knee cartilage.
   What is the function of the cartilage in the knee joint? [1]

2.7 What type of joint is the knee joint? Choose from the options below. [1]
   A Ball and socket
   B Pivot
   C Hinge
   D Gliding

2.8 Name the fluid found in the knee joint which lubricates the joint. [1]

3 **3.1** When a person runs, their heart rate increases. This is under nervous control.
   Which parts of the nervous system are involved in this process? [4]
   A The central nervous system
   B The peripheral nervous system
   C The somatic nervous system
   D The autonomic nervous system
   E The sympathetic nervous system
   F The parasympathetic nervous system

3.2 A patient with brain damage finds that she can no longer run, as she cannot co-ordinate her movements.
   Suggest **two** parts of her brain that may have been damaged. [2]

When a neurone is not conducting an impulse, it has a resting membrane potential of approximately −70 mV.

3.3 Define the term membrane potential. [1]

3.4 Explain how the resting membrane potential is maintained. [4]

The graph below shows how the membrane potential of a neurone varies during an action potential.

3.5 Explain what is meant by the term threshold potential. [2]

3.6 Indicate whether each of the following statements about the state of the neurone at position Y is true or false. [4]
   A The voltage-gated potassium channels are open.
   B The voltage-gated sodium channels are opening.
   C The outside of the membrane is becoming more positive.
   D The neurone is becoming depolarised.

3.7 Explain what is happening at position X on the graph. [4]

Unit 5 Investigating science

Introduction

Unit 5 is assessed internally by your school or college. You need to plan, carry out, analyse and make a presentation on a scientific investigation. You will be assessed on all aspects of your work. Your investigation must be carried out and written up individually. The AQA approved list of suggested (pre-approved) investigations is as follows:

+ Investigate the use of immobilised cells in bioreactors.
+ Investigate the factors that affect the efficiency of electroplating using copper.
+ Investigate how electrochemical cells work and the factors that can change the voltage output.
+ Investigate the factors that affect fermentation in the brewing industry.
+ Investigate the properties of modern shampoos.
+ Investigate the response of LDRs.
+ Investigate the factors that affect the output of a wind turbine.
+ Investigate the factors that affect reaction time.

An exemplar Student Assignment Brief, for the investigation: Investigating fermentation in the brewing industry, is available from the AQA website: https://filestore.aqa.org.uk/resources/science/AQA-17755-EX-IS.DOCX

You can choose your own investigation or one suggested by your centre, but this must be approved by AQA to confirm the suitability of the investigation and allow you to access the full range of grading criteria. This is particularly important when it comes to assessing the 'range of equipment and materials' used in assessment criterion P4.

Your investigation must also have a vocational context and be applied to a local, national or international scientific industry.

Good investigations have some of the following general features:
+ The submitted portfolio presentation is comprehensive.
    + All the assessment criteria have been addressed and met fully.
    + There is clear depth of treatment.
+ There is evidence of comprehensive research.
    + Secondary sources of information or data are used as part of the planning process and to corroborate the results of the investigation.
    + The research is clearly and fully referenced.
    + Nothing obvious has been omitted. (Your teacher/tutor will help to guide you on this.)
+ The scientific knowledge and understanding included within the presentation are clearly at Level 3 standard.
    + You should avoid submitting work that only contains GCSE standard theory or practical demand.
+ There is a wide-ranging approach with clearly identified variables or factors to investigate.
    + You should aim to investigate more than one variable or factor.
+ The investigation uses comprehensive and appropriate Level 3 practical methods and techniques.
    + A good indicator for this is the level of *accuracy* and *reliability* in the data.
    + Several *standard procedures* may have been used or considered (trialled) and rejected.
    + A range of more complex experimental techniques have been used.

Check your understanding and progress at **www.hoddereducation.co.uk/myrevisionnotes**

☐ There is a high level of analysis of the data obtained.
   ☐ There is evidence of sensible and correct manipulation of the data.
   ☐ There is a full evaluation of the data.

Owing to the nature of different practical investigations, it is not possible to give details of all the possible investigations. This chapter will give you a checklist framework to use when you are: planning your work; obtaining the data; analysing the data; and producing your portfolio presentation.

AQA are *very* specific about the meaning of some key terms. They publish a useful Glossary of terms that you should use. This is available on their website: https://www.aqa.org.uk/subjects/science/applied-general/science/teaching-resources. Key terms that are included in the AQA glossary are indicated by **_italic bold face_** text.

Prepare for a scientific investigation

Make sure you follow this checklist

You should produce a clearly identified plan including the following.

1 A statement of the tasks needed to complete your investigation and the aims/purposes/objectives of each task.
2 Details of the scientific area being studied, including any scientific theories or principles involved with your investigation.
3 A list of equipment required to carry out your investigation.
   ☐ Include the ranges, tolerances and precision of any measuring equipment used.
   ☐ Include relevant labelled diagrams.

> **Examiner tip**
>
> Ensure that any diagrams you draw use the correct scientific conventions and the correct symbols for apparatus, and are drawn in 2D. Include photos of your experimental set-up in your portfolio.

4 A statement of the **_standard procedures_**/techniques that you will need to use in your investigation.
   ☐ Include any measurements or observations to make and assess the accuracy, reliability and validity of your data and methods.
   ☐ Include all the variables/factors that you are going to investigate and their scientific rationale. You must also state the variables/factors that you are controlling, the scientific reasons why and how this will give accurate and reliable data.
   ☐ Include details of any trials carried out to assess the suitability of the standard procedures used.

> **Exam tip**
>
> Where possible use researched standard procedures. These should be fully referenced and evaluated within your bibliography. You can make alterations, adaptations and additions to researched standard procedures, but these changes must be indicated as such (see Checkpoint 6).

5 Descriptions of any trials that you have done to practise techniques or to determine the parameters of your investigation.
   ☐ Include any trial data or results obtained.
   ☐ Include details about how to use any unusual or unfamiliar pieces of apparatus or technology.

Objectives The specific points that you want to address in order to fulfil the purpose.

Precision Of a measuring instrument, the smallest division or increment that can be measured on the instrument without estimating. A ruler graduated in cm and mm can measure with a precision of measurement to ±1 mm.

Purpose The reason why you are doing an investigation; what you are trying to find out.

Range Of a measuring instrument, the maximum and minimum values that can be measured by an instrument on its current setting.

Tolerance The difference between the maximum and minimum values of the permissible errors in a measurement.

151

□ Include details about how you have determined any ranges and intervals of the data.

□ State values of any controlled variables/factors that you intend to use.

6 Stated and explained modifications to the standard procedures, techniques or scope of your investigation, made as a result of any trials that you have carried out.

□ During the planning phase you will need to carry out trials.

□ Include any alternative trialled and discarded standard procedures or techniques.

□ Fully justify your final standard procedures.

7 Research about any related commercial and industrial uses for your investigation.

□ You could include an identified, chosen context, e.g. a particular company.

□ Explain the reasons for your choice.

Your plan should appear at the front of your portfolio (following your Abstract). The plan could be written in a range of different formats, such as a Word document or a PowerPoint presentation.

> **Interval** The difference in quantity between two consecutive measurements.
>
> **Secondary data** Data that has been collected by someone else. You will probably find your secondary data on the internet.

> **Exam tip**
>
> It will help you to complete your overall plan if you keep a dated lab diary during the planning phase (and all other phases).

Carry out the investigation and record results

Make sure you follow this checklist

You need to include the following.

8 Full *risk assessments* for all your experimental procedures.

□ Include references to CLEAPPS Student Safety Sheets as appropriate.

□ You could use a similar format to the one that you used in Unit 2. Refer to the information about what to include in risk assessments on page 93.

□ Explain the control measures taken to ensure the safe use of equipment. You cannot just state what the control measures are; you must explain why they have been taken.

9 Evidence that you have correctly followed standard procedures to use a range of practical equipment and materials safely.

□ Part of this is assessed by an Observation record signed by your teacher/tutor.

□ Where possible, use a full range of equipment and materials. For example, use a *volumetric* pipette rather than a measuring cylinder to accurately measure out a quantity of liquid.

□ Include in your submitted portfolio copies of all the standard procedures you have followed. These should be suitably detailed so that someone else could follow them, collect results similar to yours and make the same conclusions as you.

10 Records of any *qualitative* data that you have used in appropriate formats.

□ Include any secondary data/information/observations that you have researched, separately to your own data and annotated to show where the information has come from.

11 Tabulated (or otherwise suitably formatted) records of any *quantitative* data.

□ Record quantitative data to a suitable (stated) level of precision (including an error judgement at the top of each measurement type).

□ Pay particular attention to the number of significant figures and decimal places that you use to record your data. They should be similar to the values used by your measuring instruments.

□ Use the correct units, in the correct format (using superscript indices, for example, $m\,s^{-1}$ *not* m/s).

> **Exam tip**
>
> CLEAPSS Student Safety Sheets are freely available on the CLEAPSS website via the link: http://science.cleapss.org.uk/Resources/Student-Safety-Sheets/

> **Exam tip**
>
> You could include photos of yourself performing the experiment(s), particularly, using more complex equipment or experimental techniques.

Check your understanding and progress at **www.hoddereducation.co.uk/myrevisionnotes**

- Include a Unit Submission form with your Portfolio, signed by you and your teacher/tutor to state that all the data recorded is yours. In some cases, you may need a partner to help you with some aspects of the experiment. This is fine, but you should include a record of this and ensure that you are the person taking the measurements.
- Any secondary data that you use, for example, corroborating data from other students, or group work, should be tabulated separately from your own data and annotated clearly to show where the data has come from.

12 An assessment of the effectiveness of the methods used to collect data.
- Assess the methods and apparatus by considering the quantitative data collected.
- Include in your portfolio a written consideration of the following points, and the impact of the effectiveness of the methods used:
  - Your data need to be complete. This means that there should not be aspects of the experiment that you have not finished, where more data could be recorded. The range of your data should be big enough to ensure that you can make a valid conclusion.
  - Your sample data size should be suitable for the experiment and enable you assess the experiment for error. You should have enough repeats for the complexity of the tasks. For example, some experiments are very complex and require considerable time to collect results. In this case, repeating the experiment a couple of times is appropriate. Other experiments are quite straightforward, and it is quick and easy to repeat them. In this case, you should repeat the experiment a few more times (e.g. five times).

    If you are using a sampling technique, for example assessing the reaction times of males versus females, you need to have enough samples (e.g. more than 20 males and an equivalent number of females).
  - The *precision* of your measurements should be appropriate for the experiment and you should consider and note the measurement error of each measured variable. For example, if you are measuring the length of a long wire, you do not need the precision of a vernier calliper or micrometer; instead, using a tape measure to measure to the nearest millimetre would be appropriate. However, to measure the diameter of a thin wire, a tape measure would be inappropriate. You should state in your table of results the precision of measurement for each measuring instrument used (e.g. ±0.001 m).
  - Identify any *anomalies* in your data. These are normally circled in a table (with a suitable key). If you have anomalies, you should keep them in your data set but repeat the reading and include the new value as well. An example table of data is shown in Table 5.1. This is from an experiment measuring the maximum electric current that can flow through fuse wires of different diameters.

Table 5.1 An example data set

Diameter, d / mm ±0.001 mm	Maximum current, I / A ±0.01 A		
	1	2	3
0.813	2.40	2.40	2.39
0.711	1.84	1.83	1.84
0.610	1.46	1.46	1.46
0.559	1.15	(1.51)	1.16
		1.15	
0.508	0.92	0.91	0.91

Key: ⬭ = anomaly

☐ Assess the *repeatability* and *reliability* of your data. If you have enough data, you can use statistical methods such as standard deviations and t-tests. Excel has these functions built-in to help you. Otherwise, you will have to examine the data by eye. Look to see if any of your data values are anomalies; if so, repeat those measurements. Determine the range of each set of readings and check to see if one set of repeats has a very different range to another set of repeats.

☐ Assess the *accuracy* of the data. This is straightforward in some cases, where you use the data to calculate a physical constant, such as the Earth's gravitational field, *g*, because you can check your value against the value given in tables of data online or in data books. In other cases, for example if you are investigating the output of electrochemical cells, you can search for values online to corroborate your values.

13 Make justified suggestions for any improvements that could be made.

☐ First review the effectiveness of the methods used to collect the quantitative data and their impact on the data. (The points considered in Checkpoint 12.)

☐ Suggest improvements that could lead to improved accuracy, reliability, repeatability and precision of measurements.

☐ Make suggestions based on modifications to the methods used and/or the equipment used.

☐ Fully explain each improvement and include suggestions on how it would improve the data.

Analyse results, draw conclusions and evaluate the investigation

Make sure you follow this checklist

You need to include the following.

14 Identification of any anomalous data, with reasons for the anomalies (refer to Checkpoint 12).

☐ Identify any anomalous data by examining the data set for values that look out of place.

☐ Repeat readings where necessary to check the anomaly and record the repeated value.

☐ Keep anomalies in your data set but identify them (e.g. by drawing a ring around the value) and include a key.

☐ Ignore anomalous data when performing calculations, such as mean averaging.

☐ Suggest reasons for any anomalies.

15 An explanation of any calculations carried out (explain the maths used).

☐ This is particularly important if you use any statistical methods to analyse your data, such as t-tests or standard deviations.

☐ Include your mathematical methods or graphical analysis in your portfolio, i.e. show your full working. If you use Excel to plot a graph and then use the graph to calculate a gradient or measure an intercept, you are advised to check any values calculated by plotting a hand-drawn graph and using that graph to draw a best-fit line and subsequently calculate a gradient.

16 Any appropriate ways of presenting your (processed and analysed) quantitative data.

☐ Include any graphs or charts that you think you need.

☐ Graphs drawn using Excel are acceptable but you must ensure such graphs are drawn to Level 3 standard and not just automatically generated by Excel.

> **Exam tip**
>
> Double check the mathematical methods that you have used by corroborating the methods online.

Check your understanding and progress at **www.hoddereducation.co.uk/myrevisionnotes**

- Graphs or charts must have:
  - a suitable scale – it is particularly important to consider the origin (0, 0) of graphs or charts drawn by Excel and amend this if the origin selected by Excel is not appropriate
  - correct gridlines
  - a suitable title
  - correctly labelled axes with correctly formatted units
  - correctly plotted points
  - correctly formatted data points or bars – points should be drawn as '+' or '×', unless there are multiple sets of data on the same graph. Make sure you have selected the correct version of bar-chart format
  - a correctly drawn best-fit line (if appropriate). Excel does a very good job of plotting straight best-fit lines but needs careful editing to plot best-fit curves. If in doubt, draw the best-fit line by hand
  - error bars (if appropriate). Excel will 'guess' values for error bars but can be edited to plot the error bars any way you want it to. The Support page on the Microsoft website can help you; use the search term 'error bars' to find the information you need.
- Add annotations (if necessary) as *text boxes* or *callouts*, straight onto the graph or chart. Callouts are particularly useful as they can point to a particular feature.
- Draw or print any charts or graphs to an appropriate size (usually, full A4 size is appropriate). These can be drawn or printed separately, or they can be integrated into the text of a Word document or PowerPoint presentation. If you are using PowerPoint, make sure you print the figures as whole slides; do not use the Handout function, otherwise the graphs and charts will be too small.

17 Clear examples (annotated screenshots) and explanations of any appropriate use of IT to process and analyse data (e.g. the use of Excel spreadsheets).
- Include screenshots showing any formulae used in Excel.
- You can annotate Excel spreadsheets using text boxes or callouts. Explanations of any functions used should be included in your portfolio.
- Include a discussion about the methods and formats used in the analysis in the context of the outcome. These should also be related to the original aims of the investigation. Justifications of the analysis methods used should also consider what further information or outcomes are derived as a result of using those methods.

18 Any valid conclusions that you have drawn.
- These need to be relevant to the purpose and objectives of the investigation, as identified in the planning stage.
- Conclusions should be based on the data obtained during the investigation.

19 An explanation of your conclusion. This should include:
- reference to any information researched from secondary sources (e.g. online references; comparison to given referenced values or data); secondary data should be researched from reliable sources of information
- reference to, and comparison of conclusions with, any researched theory, i.e. comparison with researched expected outcomes. You should make a clear distinction between conclusions based on primary data and conclusions based on secondary data.

20 An evaluation of the techniques, methods and standard procedures used and an evaluation of the results and outcomes you have obtained.

21 A full error analysis.
- This should include:
  - identification and explanation of any sources of quantitative and qualitative error
  - an assessment of qualitative errors based on the method used
  - an assessment of quantitative errors based on the spread of the data and the precision of measurement.

Exam tip

You can find advice on how to display formulae used in Excel via the Microsoft support website: **support.microsoft.com**. Use the search function to look for 'Display or hide formulas'.

Error A measurement of the uncertainty in an experiment. It can refer to a single measurement, a set of measurements, calculated values or conclusions.

My Revision Notes: AQA Applied Science Suitable for Level 3 and Level 3 Extended Certificates

☐ Discuss the source of all errors and suggest ways to minimise them. You should consider: the measuring instruments used; the number of repeats carried out; and ways of dealing with anomalies.

☐ Include an overall assessment of the percentage error in your experiment (if applicable).

Present findings of the investigation to a suitable audience

Make sure you follow this checklist

You need to include the following:

22 A combination of text and images to produce an effective and comprehensive presentation of your results for a suitable, identified audience.

☐ Integrate images within the text (e.g. Figure 1 …).

☐ Identify the audience. This may be your peers or someone who works at the company identified in your plan (if relevant). The portfolio should be appropriate for your audience.

☐ Your presentation can be submitted in a range of different (and mixed) formats, and it is essentially a *portfolio* of the work done containing all the sections listed above. You could submit: Word documents, PowerPoint presentation print-outs, leaflets or Excel spreadsheet print-outs.

☐ Ensure that any PowerPoint presentation print-outs are not too wordy. Choose the correct medium for each section that you want to present.

23 A summary of: the purpose of the investigation; the experimental data obtained; and the conclusions of the investigation (this is called an abstract).

☐ This should be put at the *beginning* of the portfolio.

24 Good correct use of appropriate scientific terminology. This must feature:

☐ the correct scientific words, conventions and symbols in context

☐ accurate spelling, punctuation and grammar

☐ *succinct* use of prose – do not make your portfolio too long; it should be comprehensive but not verbose

☐ correct signposting (referencing) of the secondary data researched.

25 A statement about the relevance of your investigation to any commercial or industrial processes identified during the planning stages.

☐ Make links between the researched science and its proposed commercial or industrial use(s).

☐ Relate the results and outcomes of your investigation to the proposed use(s).

26 A comprehensive and integrated bibliography using a Harvard Reference style.

☐ There are many different resources that can teach you how to use the Harvard Referencing system. The reference consists of an in-text component (e.g. AQA, 2020) and the full reference (AQA, 2020. AQA L3 Certificate/Extended Certificate in Applied Science – Specifications. [online] Filestore.aqa.org.uk. Available at: <https://filestore.aqa.org.uk/resources/science/specifications/AQA-1775-SP-2016.PDF> [Accessed 28 December 2020].). This element appears in an alphabetical list at the end of your report, under the title 'Bibliography'. A very useful online tool that can be used to help you to do this is available at: www.citethisforme.com/harvard/source-type.

> **Exam tip**
>
> This should be a *portfolio* submission and there is no requirement for you to write an extra 'mini-report'. Your portfolio can contain information submitted in a range of formats.

> **Exam tip**
>
> Number the pages of your final portfolio. This will help considerably when it comes to cross-referencing sections within your portfolio.

Check your understanding and progress at **www.hoddereducation.co.uk/myrevisionnotes**

- References should include the information required to locate the source (the Harvard Style reference itself), an assessment of its usefulness and a validation. For each reference you could include:
  - An assessment of usefulness – what did you use it for?
  - Validation including (as appropriate):
    - the type of publication and who published it and where (types could include: advertising/Government report or information/academic work/commercial or industrial publication/pressure group publication)
    - the purpose of the publication
    - the academic 'standing' of the author
    - information about any: peer review; editorial control; adopted textbook; book review; citations; and cross-referencing.
- An example reference using this system is shown below (Peggs and Bettin, 2008) (referencing the density of water):

Peggs, S. L. and Bettin, H., 2008. *Kaye & Laby, Tables of Physical and Chemical Constants: 2.2.1 Densities* [online] Available at: <http://web.archive.org/web/20190422110321/http://www.kayelaby.npl.co.uk/general_physics/2_2/2_2_1.html> [Accessed 29 December 2020].

Usefulness: I used the density of water to calculate the mass of water using a volume measurement. This source lists the values of the density of water at the standard pressure of 1 atm. The table gives values in intervals of 2 °C up to 100 °C. My experiment was carried out at 18 °C, so the value was easy to find.

Validation: Kaye and Laby is an online (archived) database of Tables of Physical and Chemical constants. It is the standard accepted source for constants, archived by the National Physical Laboratory (NPL), which is a government-funded National Measurement Laboratory, in conjunction with the Institute of Physics. The database is no longer updated. The purpose of the publication is to provide a comprehensive set of physical and chemical constants to the scientific community. This page was written by Sylvia Peggs and Horst Bettin, both scientists associated with the NPL. The page is a peer review of a number of references listed in the Bibliography on the page, but this data is taken from: Bettin, H. and Spieweck, F., 1990. Die Dichte des Wassers als Funktion der Temperatur nach Einführung der Internationalen Temperaturskala von 1990. *PTB-Mitteilungen*, 100, pp. 195–196.

> **Exam tip**
>
> Separate bibliographies are the best method for referencing if your portfolio uses a mixed format of different media.

6a Microbiology

The AQA Performance outcomes for this section are shown below:

Performance outcomes	Pass	Merit	Distinction
	To achieve a pass the learner must evidence that they can:	In addition to the pass criteria, to achieve a merit the learner must evidence that they can:	In addition to the pass and merit criteria, to achieve a distinction the learner must evidence that they can:
PO1 Identify the main groups of microorganisms in terms of their structure	**P1** Describe akaryotes, prokaryotes and eukaryotes in terms of their characteristic features (ultrastructure) **P2** Describe techniques used to identify microorganisms **P3** Use Gram staining techniques to identify microorganisms	**M1** Relate the characteristic features of akaryotes, prokaryotes and eukaryotes to their functions **M2** Explain how techniques used to identify microorganisms relate to the structure of the microorganisms	**D1** Compare the use of different identification techniques in biotechnological industries

Your work on this unit should be submitted as a portfolio of reports. The tasks ask you to do some practical work and some research, and the reports of each section can be produced in a number of formats: Word documents, posters, leaflets, magazine articles or PowerPoint presentation. You should choose the format that suits the report and your style the best.

For all parts of this section you must fully reference all sources of information used and integrate these references into your report (using a Harvard-style referencing system, either by bibliography or by footnotes).

The main groups of microorganisms in terms of their structure and function

REVISED ●

Make sure you follow this checklist

Task 1

Produce a report on research you have carried out about the ultrastructure of microorganisms.

Investigate the characteristic structural features of akaryotes (*viruses*), prokaryotes (*bacteria*) and eukaryotes (*microscopic fungi*). For each type, include:
- ☐ the characteristic structures and features
    - ☐ for eukaryotes, include: nucleus; cell membrane; mitochondria; endoplasmic reticulum (rough and smooth); Golgi body; ribosomes; cell wall
    - ☐ for prokaryotes, include: free DNA; plasmids; cell wall; capsule; slime coat; mesosomes
    - ☐ for akaryotes, include: nucleic acid; capsid; envelope.
- ☐ labelled diagrams
- ☐ an explanation of how each structural feature is related to function

Check your understanding and progress at **www.hoddereducation.co.uk/myrevisionnotes**

The term akaryote is confusing. It is not often used by scientists and its meaning can vary. It can refer to any cell without a nucleus (which includes all prokaryotes and some eukaryotic cells, e.g. red blood cells). Viruses are not regarded as cells but are sometimes included in the definition and sometimes not. When doing your research, it is important to be aware that AQA use the term akaryote to mean *viruses* (and *only* viruses).

> **Exam tip**
>
> You could produce a table for each type of organism showing the features, a diagram (or diagrams) of the ultrastructure and the functions of each feature within the type of organism.

Task 2

You should produce a report on experiments carried out to identify the main groups of microorganisms in terms of their structure and function.

- ☐ Investigate ways of identifying microorganisms, to include:
  - ☐ Gram staining (Gram-positive and Gram-negative bacteria)
    - ☐ Your teacher/tutor will demonstrate this technique first.
    - ☐ Identify both Gram-negative and Gram-positive bacteria.
  - ☐ microscopy (to look for characteristic features)
    - ☐ Consider both light and electron microscopy techniques.
  - ☐ colony characteristics (colour, shape, etc.).
- ☐ Relate the identification techniques to the structures of the microorganisms concerned (this relates mainly to Gram staining of bacterial cell walls).
- ☐ Compare the use of different identification techniques in biotechnological industries, e.g. environmental monitoring of water samples, testing the antimicrobial effectiveness of medical products.

Exemplar standard procedure

Gram staining of bacteria

Scope: This SP can be used to identify Gram-positive and Gram-negative bacteria.

Principle: The cell walls of Gram-positive bacteria stain purple. Those of Gram-negative bacteria stain red.

Materials needed: Crystal violet; iodine solution/Gram's iodine; decolouriser (e.g. ethanol); Safranin; bacterial culture

Equipment needed: Microscope (ideally with oil immersion objective lens); microscope slide and coverslip; wash bottle of water; Bunsen burner

Safety: A risk assessment compatible with your centre should be written or followed. Written risk assessments must be checked by a competent person before use. You must refer to CLEAPSS Student Safety Sheet 01 Microorganisms.

Procedure:

1 Apply a smear of bacteria to a slide. Air dry and then heat fix by passing it through a Bunsen burner flame a few times.

2 Add five drops of crystal violet stain to the culture. Leave for 1 minute. Rinse briefly with water and shake off excess.

3 Add five drops of iodine solution to the culture. Leave for 30 seconds, rinse briefly with water and shake off excess.

4 Rinse sample/slide with decolouriser (acetone or alcohol) for a few seconds and *rinse gently with water*. The acetone/alcohol will decolourise the sample if it is Gram negative, removing the crystal violet. Note that if the decolouriser *remains on the sample for too long, it may also decolourise Gram-positive cells*.

5 Add five drops of Safranin. Let stand for 1 minute, wash briefly with water (no more than 5 seconds) and shake off excess.

6 Examine under microscope at both ×400 and (if available) ×1000 oil immersion.

Calculations: No calculations are needed for this SP.

Expression of results: You will need evidence of completion of this task in your portfolio. This could be a series of labelled sketch observations or labelled photographs.

Using aseptic techniques to safely cultivate microorganisms

The AQA Performance outcomes for this section are shown below:

Performance outcomes	Pass	Merit	Distinction
	To achieve a pass the learner must evidence that they can:	In addition to the pass criteria, to achieve a merit the learner must evidence that they can:	In addition to the pass and merit criteria, to achieve a distinction the learner must evidence that they can:
PO2 Use aseptic techniques to safely cultivate microorganisms	**P4** Prepare risk assessments for the safe cultivation of microorganisms. **P5** Cultivate microorganisms using three different cultivation techniques. Use aseptic techniques.	**M3** Explain the control measures taken to ensure the safe cultivation of microorganisms. **M4** Explain the principles underlying the cultivation techniques used.	**D2** Evaluate the effectiveness of the aseptic and cultivation techniques used and make justified suggestions for improvement.

Make sure you follow this checklist

Tasks 3a, 3b and 3c

You should produce a report on the three different cultivation techniques that have been performed using aseptic technique. The technique should show that at least two different organisms have been used, e.g. a bacterium and a fungus.

You need to:
- ☐ know how to prepare sterile growth media (broth, agar and agar plates) for use in cultivating microorganisms
- ☐ produce risk assessments for the three different cultivation techniques used to grow microorganisms

> **Making links**
>
> You can learn more about risk assessments in Unit 2, page 93.

- ☐ know methods to safely dispose of microorganisms and equipment used when culturing microorganisms.
- ☐ grow microorganisms using three different culture techniques from the following list:
  - ☐ streak plates
  - ☐ lawn plates
  - ☐ pour plates
  - ☐ mycelial discs
  - ☐ viral plaque counts
- ☐ explain in your report the principles underlying the cultivation techniques used (e.g. culture medium used)
- ☐ use aseptic techniques
- ☐ explain in your report the control measures taken to ensure the safe cultivation of the microorganisms, including the different aspects of aseptic techniques used in the procedures
- ☐ evaluate the effectiveness of each cultivation technique used and the control measures taken, and make *justified* suggestions for improvement.

The control measures taken to ensure safe cultivation of microorganisms include the following:
- ☐ Measures that ensure there is no contamination of the culture; if such contamination occurs, you could grow unknown microorganisms that could be harmful.

> **Aseptic technique**
> A series of procedures during the culture of microorganisms which limits the chances of both the contamination of the culture medium with unwanted microorganisms, and the escape of the cultured microorganism into the air.

> **Exam tip**
>
> Note: It is unlikely that your centre will have the necessary equipment to use mycelial discs and you will certainly be unable to do viral plaque counts, which involve the use of cell cultures and infectious viruses.

Check your understanding and progress at **www.hoddereducation.co.uk/myrevisionnotes**

- ☐ Measures to ensure that the microorganisms being cultured do not escape into the air as, once the culture is established, large numbers of microorganisms will be present. This includes the proper disposal of the microorganisms and any contaminated equipment.
- ☐ Measures to ensure that if human pathogens accidentally contaminate the growth medium, they are not given ideal conditions for growth (principally, a temperature of 37 °C).

Exemplar standard procedure

Inoculation methods for the culture of microorganisms

Streak plates

This SP is a standard procedure which does not vary in different sources.

Scope: This SP is used to 'dilute' a sample of bacteria to a level where individual colonies can be seen, allowing identification of the species.

Principle: A streak plate involves the progressive dilution of a culture of bacteria or yeast over the surface of an agar plate in a Petri dish. As the streak progresses, some of the colonies on the plate grow separated from each other. The procedure should result in single, pure colonies.

Materials needed: Pure culture of bacteria in broth

Equipment needed: Inoculation loop; sterile agar plate; Bunsen burner

Safety: A risk assessment compatible with your centre should be written. Written risk assessments must be checked by a competent person before use. You should refer to CLEAPSS Student Safety Sheet 01 Microorganisms.

Procedure:

1. Sterilise the inoculation loop by passing it through a Bunsen burner flame.
2. Let the loop cool, then dip it into the broth culture containing the microorganism to be cultured.
3. Drag the loop across the surface of the agar in a zigzag motion until approximately 30% of the plate has been covered.
4. Re-sterilise the loop and turn the plate through 90 degrees.
5. Starting in the previously streaked section, drag the loop through it two to three times and continue the zigzag pattern.
6. Repeat steps 4 and 5.
7. Incubate the plate.
8. The plate should show the heaviest growth in the first section. The second section will have less growth and may have a few isolated colonies. The final section will have isolated colonies.

Calculations: No calculations needed.

Expression of results: Ensure that you label each plate with the date and the name of the cultured microorganism.

Lawn plates

This SP is based on the SP produced by CLEAPSS for lawn plates.

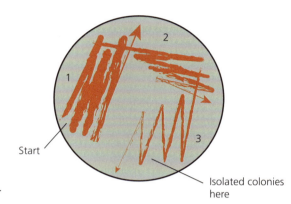

Figure 6.1 Streak plate

Scope: This SP is used to 'dilute' a sample of bacteria to a level where individual colonies can be seen, allowing identification of the species.

Principle: A lawn plate has the microbes spread on top of the agar with a sterile spreader so colonies will grow only on the surface.

Materials needed: Pure culture of bacteria in broth; beaker of Virkon disinfectant solution

Equipment needed: Sterile glass or plastic pipette; sterile agar plate; sterile spreader or sterile cotton wool bud

Safety: A risk assessment compatible with your centre should be written. Written risk assessments must be checked by a competent person before use. You should refer to CLEAPSS Student Safety Sheet 01 Microorganisms.

Procedure:

1. Using aseptic techniques suck approximately 0.3 cm³ of a liquid bacterial culture into a sterile glass or plastic pipette.
2. Re-flame the mouth of the bottle, taking care not to squeeze the teat of the pipette, and recap.
3. Using an upright sterile plate, lift one side of the lid towards the Bunsen burner to reduce contamination and squeeze in the liquid. Discard the pipette into a beaker containing disinfectant and squeeze the pipette to suck disinfectant in and out.
4. Use a sterile spreader to spread the liquid culture over the surface of the agar. Alternatively, dip a sterile cotton wool bud in liquid culture and swab it over the surface of the agar. Discard immediately into disinfectant.

Calculations: No calculations needed.

Expression of results: Ensure that you label each plate with the date and the name of the cultured microorganism.

Pour plates

This SP is based on the SP produced by CLEAPSS for pour plates. A video of the procedure in use can be seen at http://science.cleapss.org.uk/Resource-Info/Making-a-seeded-pour-plate-for-testing-anti-microbial-chemicals.aspx

Scope: This SP is usually used as a first stage in the testing of antimicrobial agents (disinfectants, antibiotics, etc.).

Principle: Sterile agar is poured over a sample of bacterial culture and then the two liquids are thoroughly mixed.

Materials needed: Pure culture of bacteria in broth; beaker of Virkon disinfectant

Equipment needed: Sterile Petri dish; sterile molten agar; sterile glass or plastic pipette

Safety: A risk assessment compatible with your centre should be written. Written risk assessments must be checked by a competent person before use. You should refer to CLEAPSS Student Safety Sheet 01 Microorganisms.

Procedure:

1 Add a small amount of liquid from a broth culture (using a sterile pipette) to the centre of a Petri dish.

2 Place the pipette in the disinfectant.

3 Pour cooled, but still molten, agar medium from a test tube or bottle into the Petri dish.

4 Rotate the dish gently, to ensure that the culture and medium are thoroughly mixed, and the medium covers the plate evenly.

5 Allow the agar to set, and then incubate the plate.

Calculations: No calculations needed.

Expression of results: Ensure that you label each plate with the date and the name of the cultured microorganism.

Mycelial discs

The mycelial disc method for culturing fungi involves placing sterile cellophane or a Petri dish culture of fungi to pick up the fungal mycelium, then cutting out discs from the cellophane and using these to 'seed' a sterile agar plate. The fungus spreads out from the disc and the diameter of the mycelium allows a measurement of growth. Certain fungi can grow in liquid broth and those can be transferred to discs of sterile filter paper instead of cellophane. The preparation of these discs is specialised and unlikely to be used in centres.

Using practical techniques to investigate factors that affect the growth of microorganisms

REVISED

The AQA Performance outcomes for this section are shown below:

Performance outcomes	Pass	Merit	Distinction
	To achieve a pass the learner must evidence that they can:	In addition to the pass criteria, to achieve a merit the learner must evidence that they can:	In addition to the pass and merit criteria, to achieve a distinction the learner must evidence that they can:
PO3 Use practical techniques to investigate the factors that affect the growth of microorganisms	**P6** Describe a range of factors that affect the growth of microorganisms.	**M5** Perform practical activities to investigate **three** factors that affect the growth of microorganisms.	**D3** Draw conclusions about how the **three** factors affect the growth of microorganisms.
	P7 Use **one** suitable technique to count/measure microorganisms. **P8** Use serial dilution techniques in **one** practical activity.	**M6** Explain the use of the technique used and perform appropriate calculations. **M7** Perform calculations to identify numbers of microorganisms in the original sample.	**D4** Evaluate the effectiveness of the measuring and counting techniques used and make justified suggestions for improvement.

Make sure you follow this checklist

Task 1

You must write a report which demonstrates knowledge and understanding of a range of factors (minimum of three) that affect the growth of microorganisms.

Check your understanding and progress at **www.hoddereducation.co.uk/myrevisionnotes**

You need to:

+ describe a range of factors that affect the growth of microorganisms.

Tasks 2a, 2b and 2c

You must investigate and report on three different factors that promote or inhibit growth from this list below, using a range of measuring techniques.

You need to:

+ be familiar with the cultivation techniques from the previous topic
+ perform practical activities to investigate three different factors that promote or inhibit growth from the following list:
  + temperature
  + pH
  + nutrients
  + aerobic/anaerobic conditions
  + antibiotics
  + antivirals
  + disinfection
  + sterilisation
  + irradiation
  + osmotic potential
  + antimicrobials (e.g. in toothpastes, mouthwash or plant derivatives such as lavender oil)
+ use one suitable technique to count or measure microorganisms. The recommended procedure is the use of a haemocytometer and light microscope (see standard procedure below), but other suitable techniques could be used by your centre. Techniques mentioned in the specification are:
  + viable counts of bacterial colonies
  + using a haemocytometer to count cells (e.g. yeast)
  + using a colorimeter to indirectly count cells
    - A colorimeter measures the penetration of light through a solution. Microorganisms in clear broth make it go cloudy, and the cloudiness measured by the colorimeter causally relates to the number of microorganisms. It can therefore be used as an indirect measurement of the population.
  + serial dilution
  + measurement of clear zones
  + viral plaque assays
    - Viral plaque assays involve the infection of a cell culture with a virus, followed by staining to identify infected cells. It is a highly specialised technique which is very unlikely to be used in centres. All other techniques are covered in the standard procedures below.

> **Osmotic potential** A measure of the tendency of a solution to withdraw water from pure water, by osmosis, across a differentially permeable membrane. Pure water has an osmotic potential of zero and the osmotic potential of a solution is always negative. The more concentrated the solution is, the more negative the osmotic potential.

> **Exam tip**
>
> Visual evidence is admissible for this task, so it is best to draw a diagram of what you see down the microscope.

Exemplar standard procedure

Making serial dilutions

This SP is adapted from the SP produced by the Royal Society of Biology and Nuffield Foundation.

Scope: This SP can be used to progressively dilute a sample (e.g. culture of microorganisms). In microbiology this is used to dilute samples where the initial number of microorganisms is too great to easily count.

Principle: A small volume of a culture is transferred into a new container and water is added to dilute the original culture. The diluted sample is then used to make further dilutions. This can be repeated several times to produce a range of concentrations.

Materials needed: Culture of microorganisms or liquid to be analysed for the presence of microorganisms; distilled water

Equipment needed: Five test tubes (number can vary depending on number of dilutions to be made); graduated pipette

Safety: A risk assessment compatible with your centre should be written or followed. Written risk assessments must be checked by a competent person before use.

Procedure:

1 Label five appropriate tubes: ×1, ×10, ×100, ×1000, ×10 000

2 Measure 11 cm^3 of your starting solution into the first tube (labelled 1).

3 Use a 10 cm^3 syringe or pipette to put 9 cm^3 of distilled water into each of the other tubes.

4 Mix the contents of the first test tube thoroughly.

163

5 Remove 1 cm³ of solution from the first tube and transfer to the tube labelled ×10.

6 Mix thoroughly.

7 Remove 1 cm³ of well-mixed solution from the ×10 tube and transfer to the ×100 tube.

8 Mix thoroughly.

9 Repeat for the ×1000 and ×10 000 dilutions.

Calculations: No calculations are needed.

Expression of results: Each test tube should be labelled and a record kept of the dilution involved.

Using a haemocytometer to measure a yeast population

This SP is a universally applied procedure.

Scope: This SP can be used to count cells (e.g. yeast) and calculate the number in a given volume of liquid.

Principle: The haemocytometer is engineered so that the volume of liquid between the haemocytometer slide and the coverslip is known. Cells are then counted in a systematic way under a light microscope and, because the volume being observed is known, the number of cells per unit volume can be calculated.

Materials needed: Culture of microorganisms or liquid to be analysed for the presence of microorganisms; 70% ethanol

Equipment needed: Haemocytometer; pipette; light microscope

Safety: A risk assessment compatible with your centre should be written or followed. Written risk assessments must be checked by a competent person before use. You should consult CLEAPSS Student Safety Sheets 01 Microorganisms and 060 Ethanol.

Procedure:

1 Clean the haemocytometer and glass cover slip with 70% ethanol.

2 Place the glass coverslip over the chambers.

3 Pipette a small quantity of the cell sample under the coverslip into the haemocytometer chamber.

4 Place the haemocytometer under a microscope and switch to the high-power objective.

5 Adjust the fine focus so that you can see the grid and count the total number of cells found in the four large corner squares and the central square (see Figure 6.2). If cells are not entirely within the square but are touching the borders, count those touching the top and right borders but do not count those touching the bottom and left borders.

Calculation: The area of the middle square and of each corner square is 1 mm × 1 mm = 1 mm²: the depth of each

square is 0.1 mm. The final volume of each square at that depth is 100 nL. The number of cells in the original sample can be calculated using the following formula:

$$\text{Total cells (cells cm}^{-3}) = \text{total cells counted} \times \frac{\text{dilution factor}}{\text{number of squares}} \times 10\,000$$

Expression of results: The number of cells per cm³ should be given in (cells cm⁻³).

Maths skills

Example calculation of number of yeast cells using a haemocytometer

The sample counted has been diluted by a factor of 10. The numbers of yeast cells counted in 5 squares on the haemocytometer were 28, 42, 39, 33 and 40. The total number of cells counted is 182.

$$\text{Total cells (cells cm}^{-3}) = \text{total cells counted} \times \frac{10}{5} \times 10\,000$$
$$= 182 \times 2 \times 10\,000$$
$$= 3\,640\,000 \text{ cells cm}^{-3}$$

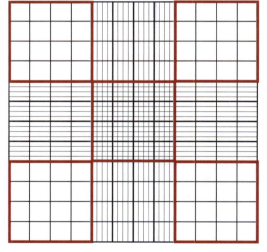

Figure 6.2 The grid pattern of an improved Neubauer haemocytometer. The red outlines indicate the squares to be used when counting yeast cells.

You need to:

+ practically investigate three factors that affect the growth of microorganisms
+ explain the use of the techniques used and perform appropriate calculations, including calculations to identify the number of microorganisms in the original sample
+ evaluate the effectiveness of the measuring and counting techniques used and make justified suggestions for improvement.

Check your understanding and progress at **www.hoddereducation.co.uk/myrevisionnotes**

Exemplar standard procedure

Investigating factors that affect the growth of microorganisms

Your centre will choose three factors from the list of 11 given in the specification. The SPs given here use the three factors which are most commonly investigated and which, together, match all the assessment criteria.

1 Investigating the effect of temperature on the growth of microorganisms

This SP is based on a suggestion in the AQA specification.

Scope: This SP investigates the effect of temperature on the growth of yeast colonies.

Principle: Yeast cells in a pour plate will grow into colonies. At a certain stage of growth, colonies become visible to the naked eye. A viable count of colonies at a set time or times therefore allows an assessment of the rate of growth. Temperature affects the activity of yeast enzymes and so can affect growth.

Materials needed: Culture of yeast; Virkon disinfectant

Equipment needed: Sterile Petri dish; sterile molten malt agar; sterile glass or plastic pipette

Safety: A risk assessment compatible with your centre should be written or followed. Written risk assessments must be checked by a competent person before use. You should consult the CLEAPSS Student Safety Sheet 01 Microorganisms.

Procedure:

1. Inoculate four sterile malt agar plates with yeast culture using the pour plate procedure and aseptic techniques as described in the previous section.

2. Place each plate in a different temperature (e.g. fridge, a cool place, room temperature, and an incubator at a temperature higher than room temperature (e.g. 45 °C)).

3. Check the plates after 24 and 48 hours. There should be individual colonies visible on all plates, but no plate should have such a dense covering that it is impossible to count individual colonies. Do a viable count of each plate (counting of all plates must be done after the same time interval). Take care not to miss colonies embedded in the agar, which will be smaller than those on the surface.

Calculation: There are no calculations for this SP.

Expression of results: Record the plate temperature and number of colonies after 24/48 hours.

> **Exam tip**
>
> AQA have criticised portfolios in which only two temperatures have been investigated. Four different temperatures is an acceptable minimum.

> **Exam tip**
>
> There is a potential inaccuracy in this technique as colony size is not considered. 50 large colonies would indicate more growth than 50 small colonies. In your evaluation, consider ways this inaccuracy could be minimised.

2 Investigating the effect of pH on the growth of microorganisms

This SP is based on a suggestion in the AQA specification.

Scope: This SP investigates the effect of pH on the growth of bacteria.

Principle: You will be provided with bacterial cultures that have been grown in media at different pH values. pH influences enzyme activity and so can affect growth. Preparing lawn plates of each bacterial culture will, after incubation, indicate the numbers of bacteria in each culture.

Materials needed: Cultures of bacteria grown in broth at different pH values (at least 4); sterile distilled water

Equipment needed: 20 test tubes; sterile graduated pipette; sterile glass or plastic pipette; sterile agar plates (one for each dilution of each pH used); sterile spreaders or sterile cotton wool buds

Safety: A risk assessment compatible with your centre should be written or followed. Written risk assessments must be checked by a competent person before use. You should consult CLEAPSS Student Safety Sheet 01 Microorganisms.

Procedure:

1. Prepare serial dilutions of each bacterial culture following the method given above, using sterile water and aseptic techniques.

2. Inoculate five sterile agar plates, each with $0.3\,cm^3$ of bacterial culture grown at each pH (one plate for each dilution) using the pour plate procedure and aseptic techniques as described in the previous section.

3. Label each Petri dish with the pH value at which the bacterial culture was grown and the dilution factor.

4. Incubate the plates for 48 hours at 25 °C.

5. Do a viable count of bacteria for each pH. Choose the strongest dilution that allows identification of individual colonies. This does not have to be the same dilution for each pH value.

Calculation: Calculate the number of bacteria in the original samples using the following equation:

$$\text{Number of bacteria (number cm}^{-3}\text{) in original culture} = \frac{\text{number of colonies}}{0.3} \times \text{dilution factor}$$

Expression of results: Record the number of bacteria in the original sample in (number cm^{-3}).

3 Investigating the effectiveness of different antibiotics on bacterial growth

This SP is based on the Royal Society of Biology/Nuffield Foundation practical on Investigating anti-microbial action.

Scope: This SP can be used to *investigate* the effects of a range of substances that may have *antimicrobial* action (e.g. antibiotics, disinfectants and personal hygiene products such as toothpaste and mouthwash).

Principle: The antimicrobial agent is soaked into a filter paper disc which is placed onto a lawn agar plate which has been freshly seeded with bacteria. The antimicrobial agent diffuses from the disc into the agar, becoming progressively less concentrated the further it diffuses. Eventually the concentration drops to a level which has no effect. After incubation, bacteria will be absent from a 'clear zone' around the disc, and the diameter of that clear zone can be used to assess the effectiveness of the antimicrobial agent.

Materials needed: Microbial broth culture – such as *Bacillus subtilis*, *Escherichia coli* or *Micrococcus luteus*; a range of different disinfectant solutions; Whatman antibiotic assay paper discs; nutrient agar

Equipment needed: Sterile Petri dishes; sterile forceps; Bunsen burner; adhesive tape; marker pen

Safety: A risk assessment compatible with your centre should be written or followed. Written risk assessments must be checked by a competent person before use. Some strains of the bacteria listed above have been associated with health hazards. You should consult the CLEAPSS Student Safety Sheet 01 Microorganisms.

Procedure:

1 Each group will need to prepare a pour plate seeded with bacteria, or have one provided. See above.

2 When the agar has set, turn the dish upside down. Divide the base into four sections by drawing a cross with the marker pen. Label the sections A, B, C, D (if using four antibiotics).

3 Using sterile forceps, place a Whatman antibiotic disc in each section; record the antibiotic in each disc.

4 Tape the lid of the Petri dish but do not seal.

5 Incubate inverted for 2–3 days at 20–25 °C.

6 Observe the plates without opening them.

7 Measure the diameter of any clear zones to compare the antimicrobial properties of the different antibiotics.

Calculation: There is no calculation needed for this SP.

Expression of results: Record clearly the diameter of each clear zone in mm next to the name of each antimicrobial agent.

The use of microorganisms in biotechnological industries

REVISED

The AQA Performance outcomes for this section are shown below:

Performance outcomes	Pass	Merit	Distinction
	To achieve a pass the learner must evidence that they can:	In addition to the pass criteria, to achieve a merit the learner must evidence that they can:	In addition to the pass and merit criteria, to achieve a distinction the learner must evidence that they can:
PO4 Identify the use of microorganisms in biotechnological industries	**P9** Describe the main features of continuous and batch processes in biotechnological industry. **P10** Describe the use of named microorganisms and the relevant industrial processes or techniques used in **two** different biotechnological industries.	**M8** Explain the benefits of an industrial fermenter. **M9** Explain the benefits to society of the use of microorganisms in the biotechnological industries described.	**D6** Compare the relevant industrial processes or techniques used for **two** named microorganisms in specific biotechnological industries. **D6** Evaluate the use of genetic engineering of microorganisms in **one** biotechnological industry.

Make sure you follow this checklist

Tasks 3a, 3b and 3c

You should produce a report on research you have carried out into the use of microorganisms in biotechnological industries.

You need to:

☐ describe the use of a range of microorganisms in biotechnological industries

☐ describe the main features of an industrial fermenter (bioreactor) and how they are used in biotechnological industries

> **Exam tip**
>
> In any account of industrial fermentation, you should explain the benefits of using an industrial fermenter over other processes.

Check your understanding and progress at **www.hoddereducation.co.uk/myrevisionnotes**

- ☐ describe industrial processes and techniques such as:
  - ☐ batch and continuous processing
  - ☐ microbial fermentation
  - ☐ immobilised enzymes
  - ☐ genetic engineering
  - ☐ biodegradation
- ☐ describe the use of microorganisms in a range of biotechnological industries, such as:
  - ☐ food production
  - ☐ environmental health
  - ☐ pharmaceuticals
  - ☐ forensic science
  - ☐ agriculture
  - ☐ alternative energies
  - ☐ waste water treatment.

To achieve a distinction, you must consider the use of genetic engineering in at least one biotechnological industry. Your account must cover advantages and disadvantages of genetic engineering, legal restrictions, public opinions (and possible misconceptions).

> **Exam tip**
>
> Your portfolio will need to cover at least two biotechnological industries, and in all cases the microorganisms used must be named. You should explain the benefits to society of the use of microorganisms in the biotechnological industries you describe.

> **Exam tip**
>
> Somewhere in your portfolio you need to compare the industrial processes used in connection with two named microorganisms in different industries. Consider culture techniques, the role of the microorganism used, the scale of production, whether batch or continuous processing is involved, etc.

6b Medical physics

The AQA Performance outcomes for this section are shown below:

Performance outcomes	Pass	Merit	Distinction
	To achieve a distinction the learner must evidence that they can:	In addition to the pass criteria, to achieve a merit the learner must evidence that they can:	In addition to the pass and merit criteria, to achieve a distinction the learner must evidence that they can:
PO1 Understand imaging methods	**P1** Describe the underlying theory behind **two** of the imaging methods listed. **P2** Select **one** medical condition and identify a suitable and an unsuitable technique for investigating the condition.	**M1** Link the underlying theory behind both of the imaging methods to explain how the images are produced. **M2** Explain why the selected technique is suitable and why the unsuitable technique selected is not appropriate.	**D1** For both methods, use calculations to support descriptions of the underlying theory.

Your work on this unit should be submitted as a portfolio of reports. The tasks ask you to do some research, and the reports of each section can be produced in a number of formats: Word documents, posters, leaflets, magazine articles or PowerPoint presentations. You should choose the format that suits the report and your style the best.

For all parts of this section you must fully reference all sources of information used and integrate these references into your report (using a Harvard-style referencing system).

Imaging methods

Make sure you follow this checklist

Task 1

You need to:

☐ describe and explain the physics behind **two** of the medical imaging methods listed below. Include any relevant calculations, and the appropriateness of the method in identifying particular medical conditions. The methods are:
  - ☐ traditional X-rays
  - ☐ X-ray digital imaging
  - ☐ CAT scans
  - ☐ PET scans
  - ☐ MRI scans
  - ☐ ultrasound
  - ☐ thermography

☐ consider, for each of your chosen imaging methods:
  - ☐ the nature and properties of waves used in the method, including how the waves fit into the electromagnetic spectrum (or sound spectrum) and how their properties relate to their speed, frequency and wavelength, and the photon energy (for electromagnetic waves). Include reference and comparison to radio waves, infrared, visible light and X-rays.
  - ☐ how the wave equation, $v = f\lambda$, relates the speed of the waves to their frequency and wavelength. Give some exemplar calculations comparing two forms of waves used in the imaging method (e.g. hard/soft X-rays; either end of the medical ultrasound spectrum).
  - ☐ (for the electromagnetic spectrum waves) how the Planck equation,

    $$E = hf \left(\text{or } E = \frac{hc}{\lambda} \right),$$ relates the energy of the wave photons to their

    frequency or wavelength. Give some exemplar calculations comparing two forms of the waves (e.g. hard X-rays and soft X-rays). The calculated energies should link to the dangers of using the technique (if appropriate).
  - ☐ how the waves are produced and *how an image/scan is produced* – include suitable diagrams to illustrate the technique, the apparatus used, and typical example images formed. Link this information back to the theory about the method.
  - ☐ why the method is suitable for imaging some parts of the body (including organs) but not others.
  - ☐ the dangers and safety precautions that need to be taken when using the waves for diagnosis and/or therapy (if appropriate). This is particularly important for the ionising radiation used in X-rays, CAT scans and PET scans.

☐ consider, if relevant to your chosen methods:
  - ☐ X-rays, CAT scans and PET scans: the factors that affect the dose of ionising radiation received by a patient
  - ☐ X-rays, CAT scans and ultrasound: the use of filters and contrast media
  - ☐ CAT scans: how 3D images are produced
  - ☐ MRI scans: how MRI scans work in terms of interaction between magnetic fields and radio waves, and the limitations of the use of MRI scans
  - ☐ ultrasound: why coupling agents are used and how they work; and calculation of the reflection coefficient between two media and its implications.

Making links

This section links to Unit 4: The human body.

Exam tip

The two most chosen techniques are traditional X-rays and ultrasound. Both of these techniques are straightforward and have large amounts of readily available information, both in books and on the internet.

Ionising Ultraviolet light, X-rays and alpha, beta and gamma rays are all ionising radiations. This means that their photons have enough energy to cause ionisation of the atoms inside living cells, causing the cell to malfunction, mutate or die.

Making links

The Institute of Physics has a useful website covering some aspects of medical physics. It can be found at: https://spark.iop.org/collections/teaching-medical-physics

Check your understanding and progress at **www.hoddereducation.co.uk/myrevisionnotes**

□ carefully select **one** medical condition and site in the body, and then identify one suitable and one unsuitable imaging technique that could be used to investigate the nature of this condition.
  □ Explain why the suitable imaging method would enable the medical staff to make a diagnosis and why the unsuitable imaging method would not be appropriate for a diagnosis.
  □ Consider the quality of the images (from both imaging methods considered) and possible dangers to the patient.

Exam tip

You will find comparing one unsuitable and one suitable technique much easier if you choose your images carefully. Ideally, pick two images (one from each technique) that show the same body part/condition.

Exam tip

The suitable and unsuitable imaging methods considered for your chosen medical condition do not have to be the ones described in the first part of this section, but it is likely that you will be able to give a better quality response if you do. If you have chosen X-rays and ultrasound there are several common medical conditions that are suitable for one method but not the other.

Radiotherapy techniques and the use of radioactive tracers

REVISED

The AQA Performance outcomes for this section are shown below:

Performance outcomes	Pass	Merit	Distinction
	To achieve a pass the learner must evidence that they can:	In addition to the pass criteria, to achieve a merit the learner must evidence that they can:	In addition to the pass and merit criteria, to achieve a distinction the learner must evidence that they can:
PO2 Understand radiotherapy techniques and the use of radioactive tracers	**P3** Describe, with the aid of diagrams, **two** radiotherapy techniques, including the disease or disorder linked with each. **P4** Identify the properties of **one** radioisotope used for a radiotherapy technique. **P5** Outline how radioisotopes can be used as tracers. **P6** Describe the dangers of radioactivity and the precautions taken to protect medical staff and patients.	**M3** Explain how each technique is used to treat specific diseases. **M4** Explain the importance of these properties. **M5** Describe the properties of **two** radioisotopes that make them suitable for use as tracers. **M6** Explain the scientific principles behind the precautions.	**D2** Discuss the invasive nature of the techniques on patients. **D3** Provide quantitative support for the explanations. **D4** Source and evaluate quantitative/graphical data of the **two** radioactive tracers.

The work that you submit in this unit will be substantially enhanced by the use of **diagrams** and **images** that link to the descriptions of the techniques chosen and the diseases/disorders chosen.

This section splits into four distinct **portfolio reports**:
a Describe and explain **two** radiotherapy techniques, including a disease/disorder linked with each.
b Identify and explain the properties of **one** radioisotope used for a radiotherapy technique.
c Outline how radioisotopes can be used as tracers and describe/evaluate the properties of **two** radioisotopes used as tracers.
d Describe and explain the dangers of radioactivity and the safety precautions taken to protect medical staff and patients.

Making links

This section links to Unit 1: Key concepts in science – Atomic structure and isotopes, page 33.

Unit 6

169

Make sure you follow this checklist

Task 2

Describe and explain two radiotherapy techniques, including a disease/disorder linked with each

- ☐ Choose **two** of the following radiotherapy techniques:
  - ☐ implants (brachytherapy)
  - ☐ external therapy
    - ☐ radioisotopes
    - ☐ gamma radiation
    - ☐ proton beam therapy (intensity modulated radiation therapy, IMRT)
    - ☐ X-rays.
- ☐ For **each** technique that you have chosen, include:
  - ☐ diagrams describing how the therapy works
  - ☐ the use of the radiotherapy technique
  - ☐ how the therapy is administered
  - ☐ how appropriate the technique is for treating diseases
  - ☐ the properties of the radioisotope used
  - ☐ the extent to which the technique is invasive to the body – explain, compare and contrast the invasive nature of each chosen therapy
  - ☐ the side effects of the radiotherapy technique.
- ☐ For your (**one**) chosen disease/disorder:
  - ☐ refer back to the diagrams used to describe how the therapy works
  - ☐ explain how **each** technique is used to treat the disease/disorder
  - ☐ explain how each chosen therapy is administered to the patient
  - ☐ state the radioisotopes (and type of electromagnetic radiation) that are chosen for each therapy and describe their properties and how they work.

> **Invasive** An invasive treatment is one in which the treatment is inserted into the body, either through an incision or via a body orifice.
>
> **Therapy** Treatment.

> **Exam tip**
>
> The most common choices include one implant and one form of external therapy.

Task 3

Identify and explain the properties of one radioisotope used for a radiotherapy technique

- ☐ Describe the nature and properties of alpha, beta and gamma radiation.
- ☐ Explain the concept of half-life.
- ☐ Explain how some radioisotopes have properties that make them more suitable for therapy than for diagnosis.
- ☐ For your (**one**) chosen radioisotope, include:
  - ☐ the name, symbol, mass number and atomic number
  - ☐ the (main) method of decay and the radioactive decay equation
  - ☐ the types of radiation emitted and their properties
  - ☐ the half-life and an explanation of this in the context of its importance in radiotherapy – consider both the physical and the biological half-life
  - ☐ the effectiveness of the radioisotope
  - ☐ the ability of the radioisotope to cause minimal damage
  - ☐ the technique used to deliver the radioisotope
  - ☐ the importance of the properties of the radioisotope and the medical context – the types of diseases/disorders that can be treated by the radioisotope/technique combination
  - ☐ the toxicity and organ affinity
  - ☐ a calculation of the time taken for the activity of the sample to fall to a level unsuitable for further use
  - ☐ a calculation of effective half-life (from the physical and biological half-lives)
  - ☐ the energies of the radioactive particles emitted by the radioisotope.

> **Dose** The dose of a radiation treatment is a measure of the total energy deposited in a small volume of tissue (the absorbed dose), the impact that the radiation has on that tissue (the equivalent dose), and the tissue's sensitivity to radiation (the effective dose).
>
> **Penetration power** The ability of a type of radiation to travel through a substance.
>
> **Toxicity** A chemical's ability to poison the body.

Outline how radioisotopes can be used as tracers and describe/evaluate the properties of two radioisotopes used as tracers

You need to:

- ☐ explain how some radioisotopes have properties that make them more suitable for use as tracers in diagnosis than for therapy

Check your understanding and progress at **www.hoddereducation.co.uk/myrevisionnotes**

- ☐ describe the uses of the radioisotope tracers in diagnosis, drug testing and research into new cures for diseases
- ☐ describe and explain how radioisotopes can be used as tracers
- ☐ explain the reasons for the widespread use of technetium-99m
- ☐ link the use of tracers to the associated medical contexts/conditions
- ☐ give a suitable range of examples – identify the common radioisotopes and give their medical uses
- ☐ explain the concept of organ affinity and the need for radiotracers to have short half-lives
- ☐ describe the type of radiation emitted by radiotracers and their detection.

For your **two** chosen radioisotopes used as tracers, you need to:
- ☐ identify the two isotopes used as tracers
- ☐ describe the medical contexts in which they are used, including: the purpose; type of illness; and location in the body
- ☐ include images/diagrams to illustrate the descriptions and explanations
- ☐ identify the properties of the radioisotopes including: the decay mechanisms and equations, the half-lives (both physical and biological), the penetration power, the toxicity, the organ affinity, the effectiveness of the radioisotope and the ability to cause minimal damage
- ☐ link the properties to the reasons why the radioisotopes are suitable for the uses described
- ☐ include radioactive decay graphs and use them to determine half-lives
- ☐ explain how the physical and biological half-lives are used to calculate the overall effective half-life of each radioisotope (including the calculations).

Task 4

Describe and explain the dangers of radioactivity and the safety precautions taken to protect medical staff and patients

You need to:
- ☐ consider the effects of ionising radiation on living tissue and the body
- ☐ explain the factors affecting the dose received and how to calculate dose equivalent
- ☐ consider the dangers of using X-rays and radioisotopes and the precautions taken to protect both medical professionals and patients
- ☐ explain how X-rays and gamma rays are different
- ☐ consider the effects of ionising radiation including the meaning of the terms: stochastic and non-stochastic; and somatic and hereditary
- ☐ consider the fact that medical uses of radioisotopes and X-rays are invasive to the body by varying degrees.

> **Exam tip**
>
> Do not include information about non-medical uses of radioactive tracers, e.g. finding leaks in pipes.

Working with radioisotopes in the laboratory   REVISED ⦿

The AQA Performance outcomes for this section are shown below:

Performance outcomes	Pass	Merit	Distinction
	To achieve a pass the learner must evidence that they can:	In addition to the pass criteria, to achieve a merit the learner must evidence that they can:	In addition to the pass and merit criteria, to achieve a distinction the learner must evidence that they can:
PO3 Demonstrate the ability to work with radioisotopes in the laboratory	**P7** Follow a standard procedure to measure the half-life of one radioisotope.	**M7** Relate the results of the experiments to the use of radioisotopes in medical treatments.	**D5** Summarise the advantages and disadvantages of alpha, beta and gamma radioisotopes in medical treatments.

In this section you have to produce a portfolio report on the following:
- ☐ An experiment to measure the half-life of **one** radioisotope.
- ☐ How your results relate to the use of radioisotopes in medical treatments.
- ☐ A summary of the advantages and disadvantages of alpha, beta and gamma radioisotopes in medical treatments.

Make sure you follow this checklist

Task 1

An experiment to measure the half-life of one radioisotope

You need to determine the half-life of **one** radioisotope in the laboratory. This needs to include a description of the experiment in full, including:

- ☐ a list of the equipment used, including the Geiger counter
- ☐ any relevant diagrams
- ☐ the safety precautions taken to ensure safety when working with radioisotopes in the school or college laboratory, including the correct handling of the radioisotopes and how these safety precautions relate to the properties of the radiation used
- ☐ the meaning and importance of background radiation, how to measure it and how to adjust your measurements to take this into account
- ☐ evidence of following a detailed standard procedure (SP) (see page 176), including:
  - ☐ your attempts at fair testing
  - ☐ your attempts at achieving accuracy
  - ☐ how you took into account the background radiation measurements
  - ☐ a copy of the SP followed
- ☐ your results (in a suitable format) including the measurements of the background radiation
- ☐ an analysis of your data including:
  - ☐ count-rates corrected for background radiation
  - ☐ a radioactive decay graph of corrected count-rate against time
- ☐ a conclusion including:
  - ☐ your measured value of the half-life of the radioisotope
  - ☐ an error value for your half-life
- ☐ an evaluation of your experiment including:
  - ☐ a comparison of your value of the half-life with the accepted value of the half-life researched online
  - ☐ a discussion of the uncertainty in this experiment including the random nature of radioactive decay
  - ☐ suggested ways to improve the experiment, including both practical suggestions and graphical techniques.

How your results relate to the use of radioisotopes in medical treatments

You need to:

- ☐ describe the nature and properties of alpha, beta and gamma radiation
- ☐ explain why the half-life of radioactive substances is important in medical treatments, including consideration of:
  - ☐ physical half-life
  - ☐ biological half-life
  - ☐ effective half-life

Check your understanding and progress at **www.hoddereducation.co.uk/myrevisionnotes**

- compare the properties (including half-lives) of a range of radioisotopes used for different medical applications (including therapy and tracers)
- compare the penetration of alpha, beta and gamma radiation in different materials and body tissues, including:
  - an account of an experiment to compare the penetration power of the different types of radiation (note: if you have carried out an experiment to measure the half-life of a radioisotope, this can be an account of a class experiment, teacher demonstration or online simulation)
  - the importance of penetrating power and its effect on the uses of different radioisotopes in medical applications (including therapy and tracers).

A summary of the advantages and disadvantages of alpha, beta and gamma radioisotopes in medical treatments

You need to:

- produce a table of the advantages and disadvantages of using alpha, beta and gamma emitting radioisotopes in medical treatments, including:
  - at least **one** example of each type of emitter
  - examples from *radiotherapy* techniques and *radiotracers*.

Exemplar standard procedure

Measuring the half-life of a radioactive substance

This SP is adapted from Practical Physics: Measuring the half-life of protactinium.

Scope: This SP can be used to measure the half-life of protactinium-234 produced using a protactinium generator.

Principle: Protactinium-234 is a decay product of the uranium decay series. It is produced from an aqueous solution of uranyl nitrate and, when mixed with an organic solvent, such as propanone, the protactinium nuclei dissolve in the organic layer and are separated from the aqueous layer. The half-life can be determined by measuring the count-rate of the organic layer using a Geiger counter over time. If your Geiger counter measures count-rate directly, use this function, otherwise record the count over 10 s.

Materials needed: Protactinium generator

Equipment needed: Clamp, boss and stand; Geiger–Muller tube; GM tube counter/rate unit; stopwatch; laboratory tray

Safety: A risk assessment compatible with your centre should be written or followed. Written risk assessments must be checked by a competent person before use. You should consult the CLEAPSS Student Safety Sheet 11 Radioactive substances. Students aged under 16 years are not permitted to handle radioactive sources, and any experiments should be demonstrated by a teacher. Students aged 16 or over in Year 12 or above (in England), can perform experiments using sealed radioactive sources under close supervision by a teacher.

Procedure:

1 Support the Geiger–Muller (GM) tube holder in a clamp, horizontally, with the window close to the upper layer; all inside a laboratory tray. This is shown in Figure 6.3. Alternatively, support the GM tube vertically resting on the top of the bottle.

2 Let the bottle stand for 10 minutes.

3 Measure and record the background count-rate. This is done with the bottle in position, as some of the background count will come from the lower layer. Repeat your measurement ten times.

4 Mix the layers thoroughly by shaking vigorously for 15 seconds.

5 Immediately place the bottle back in the tray, next to the GM tube.

6 When the two layers have separated, start the counter and start the stopwatch.

7 Record the time at the beginning of the experiment, i.e. the 'time of day' for the sample.

8 Record the count-rate every 10 seconds (or record it for 10 seconds every 30 seconds).

9 Run the experiment for five minutes (600 s) – this is a time greater than five half-lives and the count-rate should have returned to (approximately) background levels after this time.

10 Wait 2 minutes before repeating the experiment.

11 Repeat the experiment once more so that you have three sets of data in total. Examine the three data sets and decide if any of the sets are anomalous. If so, keep repeating the experiment until you have three data sets that are approximately the same.

Calculations: Calculate the mean average background count-rate. Calculate the mean average count-rate for each measurement time (0–600 s). Subtract the mean average background count-rate from each time measurement to find a mean average corrected count-rate.

Expression of results: Plot a decay graph of mean average corrected count-rate against time. Draw a suitable best fit line and use your graph to measure the half-life of protactinium-234. Repeat this measurement at different starting points along the decay graph and determine a mean average value for the half-life with an error.

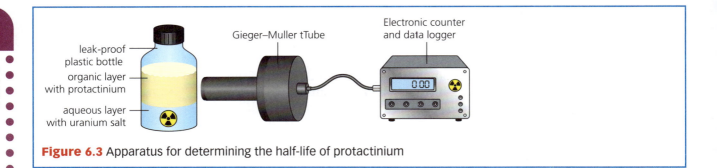

Figure 6.3 Apparatus for determining the half-life of protactinium

The medical uses of optical fibres and lasers

REVISED

The AQA Performance outcomes for this section are shown below:

Performance criteria	Pass	Merit	Distinction
	To achieve a pass the learner must evidence that they can:	In addition to the pass criteria, to achieve a merit the learner must evidence that they can:	In addition to the pass and merit criteria, to achieve a distinction the learner must evidence that they can:
PO4 Understand the medical uses of optical fibres and lasers	**P8** Describe the structure of optical fibres and how they transmit light. **P9** Follow a standard procedure to measure the refractive index of a sample of glass or Perspex.	**M8** Explain how optical fibres are used in medical treatments.	
	P10 Identify **two** medical conditions where laser light is used as a treatment.	**M9** Explain the basic scientific principles behind the use of lasers in treating both medical conditions.	**D6** Compare the advantages and disadvantages of laser and non-laser treatments for a specific medical condition.

In this section you have to produce a report on the following:
+ The structure of optical fibres and how they are used in medical treatments.
+ An experiment to measure the refractive index of a sample of glass or Perspex.
+ **Two** medical conditions where laser light is used as a treatment; the basic scientific principles of the treatments; and the advantages and disadvantages of laser and non-laser treatment for **one** specific medical condition.

Make sure you follow this checklist

Task 2

A report on the structure of optical fibres and how they are used in medical treatments

You need to:

- ☐ explain the principles behind total internal reflection and the conditions needed for this to occur
- ☐ explain the mathematical relationships between refractive index and critical angle:

$$n = \frac{\sin i}{\sin r}$$

$$n = \frac{1}{\sin c}$$

- ☐ describe the structure of optical fibres
- ☐ explain how optical fibres transmit light in terms of total internal reflection
- ☐ explain the importance of using glass with an appropriate critical angle and refractive index
- ☐ discuss why cladding is used and how the refractive index of the cladding relates to the refractive index of the core
  - ☐ give examples of the materials used to make the different layers of an optical fibre
- ☐ state and explain the different types of modes used with medical optical fibres (e.g. multi-mode v single mode)
- ☐ explain why endoscopes use glass with a high refractive index
- ☐ state **two** uses of optical fibres in medical treatments and explain how optical fibres are used in these medical treatments.

Endoscope A medical instrument using optical fibres. It can be inserted into the body to view and/or treat the internal parts of the body.

An experiment to measure the refractive index of a sample of glass or Perspex

Carry out experiments to investigate the total internal reflection of light using a sample of glass or Perspex. Use your experimental set-up to determine the critical angle for the material of the block and use this value to calculate the refractive index of the material. You need to:

- ☐ Follow a standard procedure (see below for an example) and a risk assessment for this experiment. Your centre should provide you with both of these documents and you should include copies in your portfolio for this unit.
  - ☐ Include a clear diagram of your experimental set-up.
- ☐ Produce ray diagrams showing the path of the beam of light through the block at various angles
- ☐ Vary the angle of incidence, i, of an incident beam and measure and record the angle of refraction, r, of the refracted beam. Record your data in a suitable table that you have designed yourself, including:
  - ☐ the correct tabulation of your measurements
  - ☐ repeated measurements
  - ☐ suitable precision of recording
  - ☐ the correct units
- ☐ Accurately measure the critical angle, c, and clearly calculate the refractive index, n, for a sample of glass or Perspex, using both of the formulae:

$$n = \frac{\sin i}{\sin r} \quad [1]$$

$$n = \frac{1}{\sin c} \quad [2]$$

  - ☐ You can use the values of i and the mean values of r to calculate values for n using formula 1.
  - ☐ You can use your mean value for the critical angle, c, to calculate n using formula 2.

Exam tip

Your teacher/tutor will need to write a witness statement/observation record to confirm that you have performed this experiment but a useful backup is to include photos of yourself doing the experiment.

A report on two medical conditions where laser light is used as a treatment; the basic scientific principles of the treatments; and the advantages and disadvantages of laser and non-laser treatment for one specific medical condition

You need to:

☐ identify **two** medical conditions where laser light is used as a treatment, including how laser light can be used with and without optical fibres during surgery
☐ explain why laser light is used in **each** condition
☐ state and explain how laser light is used in **each** medical treatment including, for each treatment:
   ☐ a description of the equipment used
   ☐ a description of how the treatment is administered to the patient
   ☐ an explanation of the scientific principles involved with the role of the laser light in the treatment (explain what the laser light does and how it treats the condition)
☐ explain the advantages and disadvantages (the benefits and risks) of the treatment of **one** (different) medical condition by laser and by non-laser treatment.

CLEAPSS Student Safety Sheet 12 Electromagnetic radiation is available from:

http://science.cleapss.org.uk/Resource/SSS012-Electromagnetic-radiation.pdf

---

Exemplar standard procedure

An experiment to measure the refractive index of a sample of glass or Perspex

This SP is adapted from the IOP resource TAP Episode 318: Total internal reflection.

Scope: This SP can be used to show the refraction of light through a glass/Perspex block at various angles and to determine the refractive index of the block.

Principle: When light travels from one medium to another it changes speed. The refractive index, n, is the ratio:

$$\frac{\text{speed of light in medium 1}}{\text{speed of light in medium 2}}$$

and due to the wave nature of light, for an air/glass (or air/Perspex) boundary:

$$\frac{\sin\left(\text{angle of incidence, } i\right)}{\sin\left(\text{angle of refraction, } r\right)} = \text{refractive index, } n$$

At the critical angle of the boundary, c: $n = \dfrac{1}{\sin c}$

Materials needed: Glass or Perspex rectangular block; glass or Perspex semicircular block

Equipment needed: Raybox; white A4 paper; 30 cm ruler; protractor

Safety: A risk assessment compatible with your centre should be followed. This experiment should be carried out using a raybox, and **NOT** using a laser.

Procedure:

Part a) Rectangular block

1   Place the rectangular block face down on the white A4 paper and draw around it as shown in Figure 6.4.

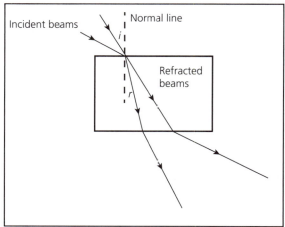

Figure 6.4 Refraction through a rectangular block

2   Remove the block and use a pencil and ruler to draw four incident beams on the paper, at roughly equal angle intervals. **Two** are shown in Figure 6.4. The incident beams should end at the air/block boundary.

3. Replace the block on the paper, turn on the raybox and adjust it (if required) to produce a single narrow ray of light.

4. Shine the beam down the first incident line and mark the position of the exit beam using a pencil and a ruler.

5. Repeat this for the other three incident lines.

6. Remove the block and join each exit beam to the entrance point, creating four refracted beams.

Part b) Semicircular block

1. Place a semicircular block on a sheet of A4 white paper and draw around it as shown in Figure 6.5.

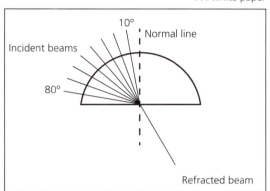

A4 white paper

Figure 6.5 Refraction through a semicircular block

2. Remove the block and draw in a normal line and incident beam lines at 10° intervals, as shown in Figure 6.5.

3. Replace the block, turn on the raybox and adjust it (if required) to produce a single narrow ray of light. Shine the beam down each incident line in turn and mark the position of the refracted beam using a ruler and a pencil. **One** refracted beam is shown for you on Figure 6.5. Turn off the raybox.

4. Remove the block and then measure and record the angle of refraction for each beam. The refracted beam will go beyond 90° at the critical angle. Record the angles of total internal reflection.

5. Replace the block and adjust the raybox beam until you have it on the critical angle. Draw in the incident ray at the critical angle and label the line c.

6. Turn off the raybox, remove the block and measure and record the angle c.

7. Repeat steps 1–6 twice more so that you have three sets of results.

8. For the angles less than the critical angle, tabulate your values of angles of incidence and angles of refraction.

Calculations: Calculate the mean angle of refraction for each angle of incidence less than the critical angle. Use

$$n = \frac{\sin(\text{angle of incidence})}{\sin(\text{mean angle of refraction})}$$

to calculate the refractive index of the block for each angle of incidence. Calculate a mean value for the refractive index.

Calculate a mean value for the critical angle. Use:

$$n = \frac{1}{\sin(\text{mean angle of } c)}$$

to determine the mean value of the refractive index. Compare your two mean values of n.

Expression of results: Put your first piece of A4 paper (from part a) into your portfolio as a record of ray diagrams showing the path of beams of light through a glass/Perspex block.

Put your subsequent pieces of A4 paper (from part b) into your portfolio, together with a table of all your results and your calculations of n.

> **Exam tip**
>
> You should create your own table for recording the data; do not use one provided for you by your centre.

6c Organic chemistry

Molecular structure, functional groups and isomerism

REVISED ◯

The AQA Performance outcomes for this section are shown below:

Performance outcomes	Pass	Merit	Distinction
	To achieve a pass the learner must evidence that they can:	In addition to the pass criteria, to achieve a merit the learner must evidence that they can:	In addition to the pass and merit criteria, to achieve a distinction the learner must evidence that they can:
PO1 Identify molecular structure, functional groups and isomerism	**P1** Outline bonding structure, nomenclature and types of formulae for compounds and functional groups.		

Performance outcomes	Pass	Merit	Distinction
	P2 Research suitable spectroscopic techniques and spectra.	**M1** Describe how infrared, NMR and mass spectra are obtained and outline the scientific principles involved.	**D1** Explain how spectra can provide specific information about structures of compounds.
	P3 Identify **one** group of compounds with a commercial or industrial use and outline their structures and uses.	**M2** For **two** compounds in the group, provide structures, skeletal formulae and identify functional groups using the correct nomenclature and scientific terminology throughout.	**D2** Explain how the structure and/or functional groups of the group of compounds make them suitable for use in medical, commercial or industrial applications.
	P4 Outline the different types of structural isomerism and geometric isomerism including suitable examples of each.	**M3** Explain the different types of isomerism with detailed examples linked to discussions of structures, shapes and molecular geometry.	**D3** Provide a detailed account of **one** compound which is biologically active, explaining the benefits and/or detrimental effects of its isomers in medical, commercial or industrial applications.
	P5 Outline optical isomerism including suitable examples from compounds found in biochemical systems.	**M4** Explain the importance of stereoisomerism in biochemical systems. Provide **one** example of a compound with specific uses, effects or actions.	

Your work on this Unit should be submitted as a portfolio of reports. The tasks ask you to do some practical work and some research, and the reports of each section can be produced in a number of formats: Word documents, posters, leaflets, magazine articles or PowerPoint presentations. You should choose the format that suits the report and your style the best.

For all parts of this section you must fully reference all sources of information used and integrate these references into your report (using a Harvard-style referencing system, either by bibliography or by footnotes).

Make sure you follow this checklist

Task 1

Produce a report on research you have carried out about the chemical bonding and structure of organic chemicals, functional groups and nomenclature.

You need to:
☐ explain carbon's ability to form strong C–C bonds due to its small atomic size, short bond length and high mean bond enthalpy
☐ explain carbon's ability to catenate and make multiple bonds
☐ correctly use the following terms:
   ☐ aliphatic
   ☐ alicyclic
   ☐ aromatic
   ☐ arene
   ☐ saturated and unsaturated
☐ know the structures of these functional groups:
   ☐ alcohol (primary, secondary and tertiary)
   ☐ aldehyde

Catenate Link together to form a connected series (of atoms, for example).

Enthalpy The sum of the internal energy, plus the product of the pressure and volume of the system.

Check your understanding and progress at **www.hoddereducation.co.uk/myrevisionnotes**

- ☐ ketone
- ☐ alkene
- ☐ halogen
- ☐ carboxylic acid
- ☐ amine (primary)
- ☐ amide
- ☐ benzene ring
- ☐ phenol
☐ use the IUPAC nomenclature for functional groups (only for aliphatic compounds, up to a maximum chain length of six carbons)
☐ use the structural (displayed) formulae for named compounds up to C_6 containing the specified functional groups
☐ interpret skeletal formulae.

In this task, you need to identify one group of compounds with a commercial/industrial use, outline their structure and uses, and give two examples of compounds in that group. For each of these examples, you should describe the structure and skeletal formulae and correctly identify functional groups. You should explain why their structure and/or functional groups make them suitable for their specified applications.

Exam tip

Make sure that, when you discuss principles and examples, you refer only to organic compounds.

Task 2

Produce a report on research you have carried out on mass spectroscopy, infrared spectroscopy and NMR spectroscopy, briefly outlining how spectra are obtained and the underlying scientific basis for the techniques.

You need to:
☐ explain how spectra are obtained and explain the science behind this
☐ explain how peaks are assigned in a spectrum
☐ be aware of spectroscopic techniques and how they support the characterisation of compounds, including structure and purity – the spectroscopic techniques could include:
- ☐ nuclear magnetic resonance
- ☐ mass spectroscopy
- ☐ infrared spectroscopy
☐ know the structures of these functional groups:
- ☐ alcohol
- ☐ aldehyde
- ☐ ketone
- ☐ alkene
- ☐ halogen
- ☐ carboxylic acid
- ☐ amine
- ☐ benzene ring
- ☐ phenol
☐ use the IUPAC nomenclature for functional groups (only for aliphatic compounds, up to a maximum chain length of six carbons)
☐ use the structural (displayed) formulae for named compounds up to C_6 containing the specified functional groups
☐ interpret skeletal formulae.

Exam tip

A useful sequence to use in your report would be: a diagram of the basic construction of each spectrometer, followed by an explanation of the type of sample used and an outline description of how the spectral 'peaks' are produced.

In this task, you need to discuss how spectra are obtained using three different techniques. You need to research and include the spectrum of a named compound to illustrate a typical output. You should assign peaks, relate them to structure and mention the effect of impurities.

Task 3

Produce a concise report containing explanations and examples of the different types of structural isomerism and stereoisomerism found in organic chemistry.

You need to:

☐ be aware of compounds which have industrial or commercial use, e.g. flavours, fragrances, lipids, pharmaceuticals, liquid crystals, terpenes, diesel fuels (including biodiesel)
☐ understand how many organic compounds can exist in isomeric forms (including a consideration of molecular shapes)
☐ understand and use the terms structural isomerism and stereoisomerism
☐ know of the following types of isomer:
  ☐ chain
  ☐ functional group
  ☐ position
  ☐ geometric and optical, including a consideration of molecular shapes and restricted rotation around carbon–carbon double bonds
☐ know definitions of:
  ☐ asymmetric carbon
  ☐ chiral centre
  ☐ optical activity
  ☐ racemic mixture
☐ understand the relationship between the three-dimensional shapes of enantiomers
☐ appreciate the importance of stereoisomerism in biochemical systems and living organisms.

> **Exam tip**
>
> Note that in this case, 'group' does not mean functional group, but a group of compounds with common uses or applications, e.g. flavours, fragrances, as listed here.

In this task, you must provide a detailed account of one compound which is biologically active, explaining the benefits and/or detrimental effects of its isomers in medical, commercial or industrial applications. You must also explain the importance of stereoisomerism in biochemical systems. Provide one example of a compound with specific uses, effects or actions.

> **Exam tip**
>
> Examples for optical isomerism should be taken from molecules found in biochemical systems, for instance lactic acid, alanine, limonene.

You need to identify one group of compounds with a commercial/industrial use, outline their structure and uses, and give two examples of compounds in that group. For each of these compounds, you should describe the structure and skeletal formulae and correctly identify functional groups. You should explain why their structure and/or functional groups make them suitable for their stated applications.

> **Exam tip**
>
> Credit is given for the use of a range of correct scientific terminology, e.g. enantiomer, racemate, optically active, etc.

Reactions of functional groups

REVISED ⬤

The AQA Performance outcomes for this section are shown below:

Performance outcomes	Pass	Merit
	To achieve a pass the learner must evidence that they can:	In addition to the pass criteria, to achieve a merit the learner must evidence that they can:
PO2 Understand reactions of functional groups	**P6** Provide examples of the reactions of **five** functional groups, stating: + reagents + conditions + observations and providing equations and explanations of the changes that occur to the functional groups.	**M5** Explain how **two** of the reactions may be used as qualitative tests for functional groups.

Make sure you follow this checklist

Task 4

Produce a report that gives examples of the different reactions of five functional groups. Include: reagents, conditions, observations, equations for the reactions and explanations of the changes that occur to the functional groups.

Check your understanding and progress at **www.hoddereducation.co.uk/myrevisionnotes**

You need to:
- [ ] understand and be able to explain types of reaction including:
  - [ ] redox
  - [ ] hydrolysis
  - [ ] esterification (condensation)
  - [ ] addition
  - [ ] elimination (e.g. dehydration)
  - [ ] enzyme-catalysed reaction
  - [ ] polymerisation
- [ ] know the reagents used to bring about change, including:
  - [ ] acidified potassium dichromate (oxidation of alcohols, aldehydes)
  - [ ] $NaOH(aq)$ or $H_2SO_4(aq)$ (hydrolysis of esters)
  - [ ] alcohols with carboxylic acids or anhydrides (esterification)
  - [ ] aqueous bromine (addition across alkene double bonds)
  - [ ] concentrated sulfuric or phosphoric acid (dehydration of alcohols)
  - [ ] Fehling's (or Benedict's) and Tollens' tests (oxidation of aldehydes)
- [ ] know the typical conditions used to carry out each type of change
- [ ] understand how functional groups change in the reactions specified above and the observations that accompany each of these changes
- [ ] provide evidence to explain how **two** of these reactions can be used as qualitative tests for functional groups.

The five functional groups you need to cover are selected from those in the previous unit:

- [ ] alcohol
- [ ] aldehyde
- [ ] ketone
- [ ] alkene
- [ ] halogen
- [ ] carboxylic acid
- [ ] amine
- [ ] amide
- [ ] benzene ring
- [ ] phenol.

The specification also mentions esterification and the hydrolysis of esters (a group not required in the previous unit). Esters are a class of compounds formed by the reaction between an alcohol and an acid.

Two of the reactions must be identified as ones which could serve as a qualitative test for the functional group. Examples include:
- alkenes reacting with bromine water
- aldehydes reacting with Benedict's, Fehling's or Tollens'
- alcohols reacting with carboxylic acids to give esters.

For this reason, you must include groups that will fulfil this requirement in your selection of five.

> **Qualitative test** A test that shows the presence or absence of a chemical but, if it is present, gives no information about the quantity.

> **Exam tip**
>
> You only need to give one example reaction from each group chosen, but it is essential that you give examples from five different groups.

Exemplar standard procedure

Bromine water test for alkenes

Scope: This SP can be used on a liquid to see if it is an alkene.

Principle: Bromine can add to alkenes because of their C=C double bond. The double bond can break allowing the addition of bromine. The orange colour of the bromine disappears from the bromine water, leaving it colourless.

Materials needed: Bromine water; test solution

Equipment needed: Test tube; test tube rack; 2 × pipettes; bung to fit test tube

Safety: A risk assessment compatible with your centre should be written. Written risk assessments must be checked by a competent person before use. Wear eye protection.

Procedure:

1 Fill a test tube about a quarter full with bromine water.

2 Add a few drops of the test liquid, place a bung in the top of the test tube and shake gently.

3 A colour change from orange to colourless indicates the test liquid is an alkene.

Calculations: No calculations are needed.

Expression of results: You should record the colour change for this reaction.

Using Benedict's solution to distinguish aldehydes from ketones

Scope: This SP can be used to distinguish aldehydes from ketones (a positive result is only given by aldehydes). The test is not a test for aldehydes as such, because other substances can give positive results.

Principle: Benedict's test is a test for reducing agents. Aldehydes reduce copper(II) ions in Benedict's solution to red copper(I) oxide.

Materials needed: Benedict's solution; test solution

Equipment needed: 250 or 500 cm³ beaker; test tube; Bunsen burner; gauze; heatproof mat; tripod; 2 × pipettes; test tube rack

Safety: A risk assessment compatible with your centre should be written. Written risk assessments must be checked by a competent person before use. You should consult the CLEAPSS Student Safety Sheets: 04 Food tests; 68 Ethanal and higher aldehydes.

Procedure:

1. Half fill the beaker with water. Place over a Bunsen burner and heat.
2. Half fill the test tube with Benedict's solution and place in the hot water.
3. Add a few drops of Benedict's solution.
4. The formation of a brick-red precipitate indicates a positive result.

Calculations: No calculations are needed.

Expression of results: You should record the colour change for this reaction.

Fehling's solution is chemically similar to Benedict's solution and the procedure and outcomes of the test are identical to those for the Benedict's test.

Esterification test for carboxylic acids

This SP is adapted from the SP produced by the Royal Society of Chemistry and Nuffield Foundation: Making esters from alcohols and acids.

Scope: This SP can be used to identify carboxylic acids, or to demonstrate the reaction between carboxylic acids and alcohols.

Principle: Alcohols react with carboxylic acids to produce esters, which have a distinctive fruity smell.

Carboxylic acid → Ester + H_2O

Materials needed: Glacial (concentrated) ethanoic acid; ethanol; concentrated sulfuric acid; 0.5 M sodium carbonate solution

Equipment needed: Eye protection; specimen tube or small test tube; test tube; 3 × plastic dropping pipettes; beaker* – 100 cm³ or 250 cm³; Bunsen burner; heat-resistant mat; tripod; gauze; crucible tongs; test tube rack; stopwatch

*Smaller beaker is for use with a specimen tube.

Safety: A risk assessment compatible with your centre should be written. Written risk assessments must be checked by a competent person before use. You should consult the relevant CLEAPSS Student Safety Sheets: 60 Ethanol; 22 Sulfuric acid; 23 Ethanoic acid. Wear eye protection and disposable gloves.

Procedure:

1. Add 1 drop of concentrated sulfuric acid to the specimen tube.
2. Add 10 drops of ethanoic acid to the sulfuric acid in the specimen tube.
3. Add 10 drops of ethanol to the mixture.
4. Put about 10 cm³ of water into the 100 cm³ beaker. Carefully lower the tube into the beaker so that it stands upright.
5. Heat the beaker gently on a tripod and gauze until the water begins to boil, then stop heating.
6. Leave the tube to stand for 1 minute in the hot water. If the mixture in the tube boils, use the tongs to lift the tube out of the water until boiling stops, then return it to the hot water.
7. After 1 minute, using tongs, carefully remove the tube and allow it to cool on the heat resistant mat.
8. When cool, pour the mixture into a test tube half-full of 0.5 M sodium carbonate solution. Mix well by pouring back into the specimen tube – repeat if necessary. A layer of ester will separate and float on top of the aqueous layer.
9. Smell the product by gently wafting the odour towards your nose with your hand.
10. A fruity smell indicates the presence of a carboxylic acid (in this case, ethanoic acid) in the sample.

Calculations: No calculations are needed.

Expression of results: You should record the observations for this reaction.

Check your understanding and progress at **www.hoddereducation.co.uk/myrevisionnotes**

Preparing organic compounds

REVISED

The AQA Performance outcomes for this section are shown below:

Performance outcomes	Pass	Merit	Distinction
	To achieve a pass the learner must evidence that they can:	In addition to the pass criteria, to achieve a merit the learner must evidence that they can:	In addition to the pass and merit criteria, to achieve a distinction the learner must evidence that they can:
PO3 Prepare organic compounds	**P7** Describe standard preparative and purification techniques used in organic chemistry, with **one** example of a preparation that uses each type of technique.	**M6** Describe how melting points and boiling points are measured and give a full description of the effects of impurities on their values.	
	P8 Carry out risk assessments and use standard procedures to prepare **two** different types of organic compounds.	**M7** Justify the choice of preparative methods including reference to yield, rate and purity or any other relevant factors.	**D4** Compare the preparative methods with those used for the industrial/commercial synthesis of the compounds.
	P9 Calculate percentage yields for each compound made and carry out practical activities to find melting or boiling points to assess purity.	**M8** Compare the differences between researched literature values and experimental values for: ✦ melting and boiling points ✦ yield obtained.	**D5** For **one** of the compounds prepared/extracted, choose a suitable spectroscopic technique and provide a detailed explanation of how it is used to assess purity and characterise the compound.
	P10 Produce a report on each preparation, describing the methodology, equipment used and outcomes for each.	**M9** Draw conclusions which are linked to the yields obtained and levels of purity achieved for each organic compound.	**D6** Suggest improvements to increase the yield and purity of the compounds.

Make sure you follow this checklist

Task 1

Produce a report that exemplifies the four standard practical techniques available to synthesise or extract organic compounds and the purification techniques stipulated in the unit content. Include diagrams of the apparatus used in each case, with a short explanation and examples of the applications of the different techniques.

You need to:
☐ describe the basic preparative techniques including:
  ☐ reflux to heat volatile compounds safely and without loss of material
  ☐ distillation to separate a volatile liquid from a mixture
  ☐ fractional distillation
  ☐ steam and water distillation to extract relatively volatile water-immiscible components from a natural material
☐ give examples of preparations which use each technique, including industrial/commercial synthesis of compounds.

183

Task 2

Apply two techniques to prepare or extract two organic compounds from a fragrance or flavouring, a common pharmaceutical, a dye, solvent, biofuel, etc. Include a risk assessment, carry out a standard procedure and refer to yield and purity to indicate successful outcomes.

You need to:
- [ ] describe how reaction mixtures can be separated and products purified, including:
  - [ ] precipitation, filtration under reduced pressure and recrystallisation
  - [ ] the use of a separating funnel for washing and separating immiscible liquids
  - [ ] final re-distillation or fractional distillation
- [ ] give examples of preparations which use each of these techniques including industrial/commercial applications
- [ ] explain that melting point and boiling point are criteria for purity
- [ ] explain how the presence of impurities may affect values of melting point and boiling point (values, range, **sharpness**)
- [ ] describe how melting and boiling points are measured
- [ ] explain how the choice of method for synthesis is based on:
  - [ ] yield
  - [ ] rate
  - [ ] purity of the product
  - [ ] safety.

> **Exam tip**
>
> You must assess the purity of any product made by *measuring* the boiling point. Simply quoting the distillation temperature is not sufficient.

> **Exam tip**
>
> When explaining the effect of impurities on melting point, make sure you cover all three of values, range and sharpness.

> **Sharpness** When related to melting and boiling points, this is the precision and consistency of the melting/boiling point. Lack of sharpness leads to a range of temperatures within which the substance may melt or boil.

Task 3

Using your notes from the practical preparations, produce two reports, one for each of two of the preparations, describing the methodology, equipment and outcome from each preparation.

In your reports of the preparations/extraction, you need to:
- [ ] identify the compound (including its structure) and show evidence of research into possible methods for its synthesis or extraction
- [ ] describe the procedure chosen (to include both the preparation/extraction stage and all purification stages) and justify its use
- [ ] describe the type of reaction (or, for an extraction, the theory behind the method including any relevant formulae and stoichiometry)
- [ ] identify any hazards and include a full risk assessment
- [ ] explain how to record measurements
- [ ] record tabulated data (masses, volumes as relevant, and melting points or boiling points)
- [ ] calculate percentage yields (for preparations)
- [ ] evaluate your methodology (including an evaluation of the yields and levels of purity obtained) and suggest modifications that may improve yields or increase purity
- [ ] explain how infrared, NMR and mass spectra can be used to characterise the compounds prepared or extracted
- [ ] apply, for **one** of the two compounds, a suitable spectroscopic technique using library spectra (available online). Discussion should centre on the use of the spectra to characterise the compound and how it can identify levels of impurity in the sample. The use of spectra for precursors and likely side products may also be relevant.

> **Making links**
>
> You can learn more about writing risk assessments in Unit 2, page 93.

> **Exam tip**
>
> Issues you could consider as possible improvements (although they will vary with the procedure carried out) include: reaction (reflux) time; steam distillation times; loss of product during reflux or distillation; using an excess of reagents to improve yields; catalysts; reversible reactions; alternative reagents; purification stages; loss of product; removal of impurities; drying stages; alternative processes.

Check your understanding and progress at **www.hoddereducation.co.uk/myrevisionnotes**

Two SPs are presented below, one *preparation* and one *extraction*, but you will carry out preparations or extractions decided by your centre, which may be different from these.

Exemplar standard procedure

Preparation of ethyl ethanoate

Scope: This SP is specific to the preparation of ethyl ethanoate.

Principle: This experiment is a quantitative esterification reaction. Ethanoic acid, ethanol and sulfuric acid catalyst are heated to form ethyl ethanoate. The ester is then isolated and collected by distillation.

Materials needed: Glacial ethanoic acid; ethanol; concentrated sulfuric acid

Equipment needed: $3 \times 10\,cm^3$ measuring cylinders; Quickfit $100\,cm^3$ round-bottomed flask; reflux/Liebig condenser; stillhead; receiver adapter; thermometer pocket; bosses and clamps; retort stands; anti-bumping granules; spatula; electric heating mantle; $0-100\,°C$ thermometer; $50\,cm^3$ conical flask and stopper; electronic balance.

Safety: A risk assessment compatible with your centre should be written or followed. Written risk assessments must be checked by a competent person before use. Techniques should be performed in a fume cupboard. You should consult the relevant CLEAPSS Student Safety Sheets: 22 Sulfuric acid; 23 Ethanoic acid; 60 Ethanol. You should wear suitable goggles, a lab coat and use suitable safety gloves for handling the organic solvents. The addition of concentrated sulfuric acid to ethanol generates a lot of heat, so add slowly and carefully. Glacial ethanoic acid is an irritant and can cause burns. Handle concentrated sulfuric acid only when wearing suitable chemical-handling gauntlets.

Procedure:

1. Place a dry $10\,cm^3$ measuring cylinder onto a balance and tare the balance. Add $10\,cm^3$ of ethanol and measure and record the mass of ethanol added.

2. Wearing suitable protective gloves, pour $10\,cm^3$ of glacial ethanoic acid and $10\,cm^3$ of ethanol into a Quickfit $100\,cm^3$ round-bottomed flask. Using chemical-handling gauntlets, very carefully and slowly, add $5\,cm^3$ of concentrated sulfuric acid, swirling the contents as you do so. This will generate heat.

3. Using a spatula, add some anti-bumping granules, and set up the apparatus for reflux heating, using the reflux/Liebig condenser and the electric heating mantle.

4. Gently heat the mixture for about 10 minutes on a low heat, using the heating mantle, ensuring that the temperature is kept below the boiling point of the mixture. Any ethyl ethanoate produced will condense on the inside of the reflux/Liebig condenser and drip back down into the flask.

5. Turn off the heating mantle and allow the apparatus to cool. Reconnect the apparatus into a distillation configuration, using the stillhead, thermometer pocket, thermometer, reflux/Liebig condenser, receiver adaptor and conical flask.

6. Gently heat the mixture so that the temperature of the thermometer is kept just above $77\,°C$, which is the boiling point of ethyl ethanoate. Collect the ethyl ethanoate fraction produced just above $77\,°C$. Keep heating until all the ethyl ethanoate has left the mixture and has dripped into the conical flask.

7. Place a stopper on the conical flask.

Calculations: No calculations required.

Expression of results: Label the conical flask as ethyl ethanoate and record the mass of ethanol used in the preparation.

Purification of the ethyl ethanoate

Scope: This is a procedure for the purification a solution of ethyl ethanoate distillate produced by the Preparation of ethyl ethanoate SP.

Principle: Washing the ethyl ethanoate distillate with saturated sodium carbonate solution removes any acidic impurities, such as any sulfuric acid catalyst from the solution. The addition of calcium chloride solution to the ethyl ethanoate distillate causes a reaction with any alcohol impurities, forming hydrates, and so removes any unreacted ethanol. The addition of solid calcium chloride removes any water that has been evaporated off the mixture when it was distilling, (also forming a series of hydrates).

Materials needed: Ethyl ethanoate distillate from 'Preparation of ethyl ethanoate' SP; $20\,cm^3$ of saturated sodium carbonate solution; $10\,cm^3$ of saturated calcium chloride solution; $5\,g$ of anhydrous calcium chloride granules.

Equipment needed: $50\,cm^3$ separating funnel with stopper; $4 \times 50\,cm^3$ conical flasks; conical flask stopper; stand, boss and clamp; spatula; filter funnel; filter paper; Quickfit distillation apparatus (as in 'Preparation of ethyl ethanoate' SP); electronic balance.

Method:

1. Pour the ethyl ethanoate distillate into a separating funnel and add $20\,cm^3$ of saturated sodium carbonate solution. Add the stopper and shake well – taking care to lift the stopper after each shake to allow any carbon dioxide gas to escape. Once all the effervescence has stopped, allow the mixture to settle out into two layers.

2. Use the tap on the separating funnel to discard the lower aqueous layer.

3. Add $10\,cm^3$ of saturated calcium chloride solution to the ethyl ethanoate distillate in the separating funnel. Add the stopper and shake well. Allow the mixture to settle out into two layers.

4 Use the tap on the separating funnel to discard the lower aqueous layer.

5 Run off the treated ethyl ethanoate solution into a 50 cm³ conical flask, add 5 g of anhydrous calcium chloride granules using a spatula, stopper the flask and shake well. Allow the solution to stand for 20 minutes.

6 Filter the ethyl ethanoate solution to remove the residue, allowing the filtrate (the remaining purified ethyl ethanoate) to run into a 50 cm³ conical flask. Replace the stopper.

7 Weigh and record the mass of a dry 50 cm³ conical flask and stopper.

8 Use the distillation apparatus shown in the 'Preparation of ethyl ethanoate' SP to gently distil all the purified ethyl ethanoate into the weighed conical flask, measuring and recording the boiling point of the purified ethyl ethanoate as you do so. Place the stopper on the conical flask.

9 Weigh the mass of the conical flask, stopper and distilled ethyl ethanoate.

Calculations: Calculate the mass of distilled ethyl ethanoate produced and use the mass of ethanol recorded in the 'Preparation of ethyl ethanoate' SP to determine the yield of this preparation and purification.

Expression of results: Express your calculated yield as a percentage. The theoretical yield for this experiment is about 67% as the reaction between the excess ethanoic acid and the limiting ethanol is a reversible reaction and approximately two-thirds of the reactants are converted to the ester product. Compare your calculated yield to the theoretical yield.

Exemplar standard procedure

Extraction of limonene from orange peel

Scope: This SP is specific to the extraction of limonene.

Principle: Limonene is an oil found in the peel of citrus fruit. When water is added to orange zest, the limonene mixes with the water but does not dissolve. Limonene boils at 74 °C and can be extracted by distillation. Although water will also be present in the distillate, the limonene separates from it and forms a layer floating on top of the water.

Materials needed: Oranges; distilled water

Equipment needed: Citrus zester; 2 × 100 cm³ measuring cylinder; anti-bumping granules; round-bottomed flask; glass rod; funnel; distillation apparatus (Liebig condenser); Bunsen burner; gauze; heatproof mat; dropping pipette

Safety: A risk assessment compatible with your centre should be written or followed. Written risk assessments must be checked by a competent person before use. The chemicals in this technique are all low hazard.

Procedure:

1 Collect the zest of two oranges using the citrus zester.

2 Add anti-bumping granules to the round-bottomed flask.

3 Add the orange zest to the round-bottomed flask, using a funnel and pushing the zest through the funnel with a glass rod.

4 Add 50 cm³ distilled water to the flask.

5 Set up the flask with a Liebig condenser as shown in Figure 6.6.

6 Collect the distillate. The mixture of distillate and water boils at around 90 °C. Stop when you have about 30 cm³ of distillate.

7 If necessary, the limonene can be separated from the water using a dropping pipette.

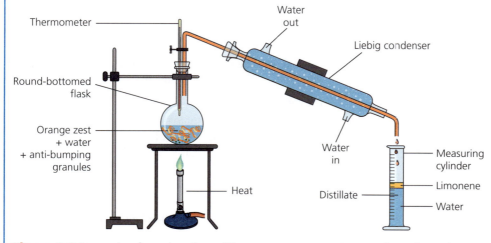

Figure 6.6 Apparatus for extraction of limonene

Calculations: No calculations are needed.

Expression of results: You need to label your completed samples, including your name (initials) and the date.

Check your understanding and progress at **www.hoddereducation.co.uk/myrevisionnotes**

Exemplar standard procedure

Preparation of aspirin

Scope: This SP is specific to the preparation of aspirin.

Principle: Aspirin is the common name for acetylsalicylic acid. It is prepared by the reaction of salicylic acid with ethanoic anhydride. Sulfuric acid is used as a catalyst.

Materials needed: Salicylic acid; ethanoic anhydride solution; concentrated sulfuric acid; distilled water

Equipment needed: $3 \times 250\,cm^3$ beakers; $250\,cm^3$ measuring cylinder; $10\,cm^3$ measuring cylinder; Bunsen burner; tripod; gauze; heatproof mat; reflux condenser; $50\,cm^3$ round-bottomed flask; funnel; ice bath; Buchner apparatus

Safety: A risk assessment compatible with your centre should be written or followed. Written risk assessments must be checked by a competent person before use. You should consult the relevant CLEAPSS Student Safety Sheets: 22 Sulfuric acid; 26 Salicylic acid. Salicylic acid is an irritant; sulfuric acid is corrosive; ethanoic anhydride is flammable.

Procedure:

1 Place a $250\,cm^3$ beaker, half-filled with water, on top of a tripod and gauze.

2 Clamp a $50\,cm^3$ round-bottomed flask so that it is partially submerged in the beaker of water and fit a reflux condenser to it vertically. Run water through the condenser.

3 Add anti-bumping granules to the flask.

4 Add 2.0g of salicylic acid, $5\,cm^3$ of ethanoic anhydride, and five drops of concentrated sulfuric acid, separately, to the flask through a funnel.

5 Heat the water in the beaker and reflux the mixture for 30 minutes.

6 During this time, place $150\,cm^3$ of distilled water, in a $250\,cm^3$ beaker, into an ice bath.

Isolation of aspirin from the reaction mixture

1 Allow the reaction mixture to cool down for a few minutes.

2 Pour the reaction mixture into the cold water. Stir for a few minutes and allow the mixture to settle.

3 Filter the mixture using Buchner apparatus. Wash any residue remaining in the beaker into the funnel with more distilled water.

4 Scrape the solid residue from the filter paper into another beaker.

Purification of the aspirin

Recrystallise the aspirin using a small quantity of boiling water.

Glossary

Accuracy An accurate result is judged to be close to the true value.

Anomaly A value that deviates from what is standard, normal or expected.

Aqueous medium A liquid which contains water, i.e. water or an aqueous solution.

Aseptic technique A series of procedures during the culture of microorganisms which limits the chances of both the contamination of the culture medium with unwanted microorganisms, and the escape of the cultured microorganism into the air.

Assimilation The processes by which an organism incorporates absorbed nutrients into the body.

Baroreceptor Sense organ which detects pressure.

Bias A source may show a bias where it shows an inclination or a prejudice for or against a viewpoint, person or group.

Block An area of the Periodic Table where all the elements have their highest energy level electrons in the same orbital type.

Catalyst A substance which speeds up a chemical reaction and is left unchanged at the end of that reaction.

Catenate Link together to form a connected series (of atoms, for example).

Chemoreceptor Sense organ which detects a chemical.

Concentration gradient The difference in the concentration of a solute between two areas. The bigger the difference, the 'steeper' the concentration gradient.

Conduction The transfer of thermal energy from hot to cold by the vibration of particles through a solid or a liquid.

Convection The transfer of thermal energy by the translation (movement) of particles from somewhere hot to somewhere cold.

Corroboration Using evidence which confirms or supports a scientific statement, theory or finding.

Displacement Of an object, its distance away from an origin in a given direction.

DNA Deoxyribonucleic acid; a self-replicating nucleic acid which is found in nearly all living organisms and provides a chemical code for the formation of proteins.

Dose The dose of a radiation treatment is a measure of the total energy deposited in a small volume of tissue (the absorbed dose), the impact that the radiation has on that tissue (the equivalent dose), and the tissue's sensitivity to radiation (the effective dose).

Efficiency The ability of a device to transfer energy input into useful energy output. Usually expressed as a percentage, %.

Electrical characteristic A voltage–current (VI) graph, with current plotted on the x-axis and voltage plotted on the y-axis.

Electrochemical gradient A situation where there is a gradient of electrical charge across a membrane.

Emission spectrum The collection of (unique) different coloured photons emitted by a hot element.

Endoscope A medical instrument using optical fibres. It can be inserted into the body to view and/or treat the internal parts of the body.

Enthalpy The sum of the internal energy, plus the product of the pressure and volume of the system.

Enzyme A protein molecule which catalyses a chemical reaction in the body.

Epithelium The outermost layer of cells in animal tissues.

Error A measurement of the uncertainty in an experiment. It can refer to a single measurement, a set of measurements, calculated values or conclusions.

Eukaryotic Cells which contain membrane-bound organelles and a nucleus.

Extrinsic protein A protein attached to the outside or inside of the lipid bilayer in a cell membrane.

Food chain The sequence of transfers of energy (in the form of food) from organism to organism.

Food web The interconnection of food chains in an ecosystem.

Gradient The slope of a line on a graph (vertical interval ÷ horizontal interval).

Homeostasis The maintenance of a constant internal state within the body.

Hypothalamus An area in the floor of the brain that maintains the body's internal balance, often by stimulating the release of hormones from the pituitary gland.

Inertia The property of a moving object that opposes a change in its motion. Inertia is linked to momentum.

Interval The difference in quantity between two consecutive measurements.

Intrinsic protein A protein embedded in the lipid bilayer of the cell membrane, sometimes completely penetrating it.

Invasive An invasive treatment is one in which the treatment is inserted into the body, either through an incision or via a body orifice.

Check your understanding and progress at **www.hoddereducation.co.uk/myrevisionnotes**

Ion A charged particle formed when an atom loses or gains electrons. Some ions are formed from a small collection of atoms, such as sulfate or carbonate ions.

Ionising Ultraviolet light, X-rays and alpha, beta and gamma rays are all ionising radiations. This means that their photons have enough energy to cause ionisation of the atoms inside living cells, causing the cells to malfunction, mutate or die.

Lymphatic system Part of the immune system, which also maintains fluid balance and plays a role in absorbing fats and fat-soluble nutrients. It consists of a network of vessels that run through the whole body and drain into the blood system at the subclavian veins in the neck.

Microvilli Microscopic projections of a cell membrane.

Myofibril A contractile fibre in skeletal muscle.

Myofilament Protein filament found in a myofibril. There are two types: actin and myosin.

Negative feedback A process where a change causes a series of events which reverse that change.

Objectives The specific points that you want to address in order to fulfil the purpose.

Organelle Structure found inside a cell which has a specific function.

Osmotic potential A measure of the tendency of a solution to withdraw water from pure water, by osmosis, across a differentially permeable membrane. Pure water has an osmotic potential of zero and the osmotic potential of a solution is always negative. The more concentrated the solution is, the more negative the osmotic potential.

Ossification The process of bone formation.

Oxidation A reaction involving the addition of oxygen, the removal of hydrogen or the loss of electrons. The opposite of oxidation is reduction.

Parallel circuit A circuit where two or more components are connected to the same points in the circuit with junctions.

Penetration power The ability of a type of radiation to travel through a substance.

Peristalsis Rhythmic muscular contractions of the digestive tract (oesophagus, stomach and intestines). Peristalsis moves food through the gut.

Pituitary gland A gland hanging from the floor of the brain which produces hormones which control the activity of endocrine (hormone-producing) glands around the body.

Polar Some overall neutral molecules can be described as polar molecules. The shared electrons in the bonds tend to be closer to one of the atoms than to the others. As a result, one end of the molecule becomes slightly negatively charged and the other end becomes slightly positively charged.

Polypeptide A chain of amino acids of insufficient length to be called a protein.

Precise Precise measurements show very little spread about the mean value. Precision depends only on the extent of random errors – it gives no indication of how close results are to the true value. It can be expressed numerically by measures of imprecision (e.g. standard deviation).

Precision Of a measuring instrument, the smallest division or increment that can be measured on the instrument without estimating. A ruler graduated in cm and mm can measure with a precision of measurement to ±1 mm.

Prokaryotic Prokaryotic cells are simple cells that do not have a true nucleus or other membrane-bound organelles.

Purpose The reason why you are doing an investigation; what you are trying to find out.

Qualitative test A test that shows the presence or absence of a chemical but, if it is present, gives no information about the quantity.

Radiation The transfer of thermal energy by the emission of infrared electromagnetic radiation from hot objects to colder surroundings.

Range Of a measuring instrument, the maximum and minimum values that can be measured by an instrument on its current setting.

Reliability The extent to which an experiment, test or measuring procedure yields consistent results on repeated trials. A measure is said to have a high reliability if it produces similar results under consistent conditions.

Repeatability The precision obtained when measurement results are produced in one laboratory, by a single operator, using the same equipment under the same conditions, over a short timescale. A measurement is 'repeatable' in quality when repetition under the same conditions gives the same or similar results, e.g. when comparing results from the same learner or group using the same method and equipment.

Reproducibility The precision obtained when measurement results are produced by different laboratories (and therefore by different operators using different pieces of equipment). A measurement is 'reproducible' in quality when reproducing it under equivalent (but not identical) conditions gives the same or similar results from different learner groups, methods or equipment – a harder test of the quality of data.

Resorption The absorption into the circulation of material from cells or tissue.

RNA Ribonucleic acid; a single-stranded nucleic acid which plays a role in protein synthesis.

Sarcomere The structural unit of skeletal muscle.

Secondary data Data that has been collected by someone else. You will probably find your secondary data on the internet.

189

Secretion The release of a substance from the inside of a cell to the outside.

Sensationalist A source is said to be sensationalist if it presents a scientific story in a way that is intended to provoke public interest or excitement at the expense of scientific accuracy.

Series circuit A circuit where the components are connected in a complete loop, one after another.

Sharpness When related to melting and boiling points, this is the precision and consistency of the melting / boiling point. Lack of sharpness leads to a range of temperatures within which the substance may melt or boil.

Sub-shell One of the orbitals that makes up an electron energy shell (energy level).

Substrate-linked phosphorylation The formation of ATP which occurs when a reaction in the cell produces enough energy to convert ADP to ATP, without the involvement of the electron transfer chain.

Therapy Treatment.

Threshold potential The membrane potential which, when reached, initiates a nerve impulse.

Titration A volumetric way of carrying out a neutralisation reaction. A volumetric pipette measures out a precise volume of acid/alkali and this is titrated against an alkali/acid using a burette and an indicator to find the end point. (Volumetric means 'measured volume'.)

Tolerance The difference between the maximum and minimum values of the permissible errors in a measurement.

Toxicity A chemical's ability to poison the body.

Trophic level The organisms in an ecosystem which occupy the same level in a food chain.

Ultrafiltration Filtration of small molecules.

Ultrastructure The fine structure of cells which is only visible with electron microscopes.

Validity The suitability of an investigative procedure to answer the question being asked.

Volatility How readily a substance vaporises. Volatility depends on temperature and pressure, but a substance with a high volatility is more likely to exist as a vapour (gas), whereas a substance with a low volatility is more likely to be a liquid or solid.

Voltage-gated channel A channel protein in the cell membrane which is only open at a limited range of membrane potentials.

Check your understanding and progress at **www.hoddereducation.co.uk/myrevisionnotes**

Answers to Exam practice questions can be found at www.hoddereducation.co.uk/myrevisionnotesdownloads

Unit 1

REVISED ●

Biology

Now test yourself

1 Mesosome and capsule
2 They are made of different materials (cellulose in plant cells, peptidoglycan in bacteria).
3 **Similarity**: both make proteins; similar structure. **Difference**: different size (80S in eukaryotes, 70S in prokaryotes.)
4 (Pentose) sugar, phosphate, nitrogenous base
5 The sugar in DNA is deoxyribose, in RNA it is ribose; DNA contains thymine instead of uracil.

Maths skills practice questions

1 Real length $= \dfrac{\text{length on photo}}{\text{magnification}}$

$= \dfrac{20\,000}{3000}$

$= 6.67\,\mu m$

2 The size of the photograph is twice as large as the original image in the electron micrograph.

Now test yourself

6 Proteins and phospholipids
7 Lipid-soluble chemicals can get through the lipid bilayer, water-soluble chemicals can only get through protein channels or carriers. There is a much higher area of lipid bilayer than protein on the cell membrane.
8 Cells live in an aqueous medium. The optimal way for the hydrophobic tails to avoid water and for the hydrophilic heads to be in water is a double layer (bilayer) with the hydrophilic heads facing outwards.
9 Simple diffusion does not require a specific structure to pass through. Facilitated diffusion does.
10 Active transport requires energy, facilitated diffusion does not; facilitated diffusion carries substances with (down) a concentration gradient, active transport carries substances against (up) a concentration gradient.
11 A molecule on the surface of a membrane that allows recognition of the cell by the immune system.
12 To ensure one-way flow of blood
13 Myogenic contractions do not require any nervous stimulation, normal contractions do.
14 B C A D
15 The sympathetic system
16 The sinoatrial node

17 In the blood
18 By controlling / increasing the permeability of the collecting duct walls to water
19 Glucagon
20 Type 2
21 Carbohydrates are converted to sugar in the body, so eating any carbohydrate can raise blood sugar levels.
22 Excretion and osmoregulation
23 The filter in the Bowman's capsule only allows small molecules to pass through. Glucose is a small molecule, but proteins are large molecules so do not pass through.
24 The adrenal cortex
25 Vital capacity is the maximum amount of air that can be breathed in or out in a single breath, whereas tidal volume is the amount of air breathed in or out during normal breathing.
26 Glycolysis
27 Phosphorylation
28 Pyruvate / pyruvic acid
29 It is used as a supply of electrons.
30 The inner membrane
31 Oxygen is the final electron acceptor for the electrons.
32 Because the measurement is not a direct measurement of heat generated
33 ATP, reduced NADP (NADPH), oxygen
34 Chloroplast
35 It needs ATP and reduced NADP, which are produced in the light-dependent stage.
36 Nitrate (accept nitrogen)
37 Low temperatures slow down the reactions of photosynthesis.
38 The organisms use some of the energy in GPP, before it can be passed on as NPP.
39 Plants contain a lot of cellulose, which most animals cannot digest very successfully.
40 1st consumers will receive 10% $\left(\dfrac{1}{10} \right)$ of 21 564 kJ

$= \dfrac{21\,564}{10} = 2156.4\,kJ$

2nd consumers will receive 10% of 2156.4 kJ = **215.64 kJ**
41 The water content of organisms varies over time and is not living tissue.
42 At each trophic level energy is lost rather than being passed to the next trophic level. If we eat plants, there is only one stage where energy is lost. If we eat animals, two or three stages are involved.
43 They may not receive all the amino acids that they need, as no single plant contains all the necessary amino acids.

Chemistry

Now test yourself

1

Atom/ion name	$^A_Z X$ notation	Number of protons	Number of neutrons	Number of electrons
Hydrogen-3	$^3_1 H$	1	2	1
Fluorine-19	$^{19}_9 F$	9	10	9
Sodium-23$^+$ (ion)	$^{23}_{11} Na^+$	11	12	10
Sulfur-32^{2-} (ion)	$^{32}_{16} S^{2-}$	16	16	18

2 All three (isotopes) have the same number of protons (1), but different numbers of neutrons (0, 1, 2).

3 $^{32}_{16}S$; $^{33}_{16}S$; $^{34}_{16}S$; $^{36}_{16}S$

4 A sub-shell is an orbital or a combination of orbitals, which describes the places with the highest probability of finding electrons.

5 One 2s and three 2p.

6

Atom	Sub-shell electron configuration
Hydrogen	$1s^1$
Oxygen	$1s^2\, 2s^2\, 2p^4$
Calcium	$1s^2\, 2s^2\, 2p^6\, 3s^2\, 3p^6\, 4s^2$
Nickel	$1s^2\, 2s^2\, 2p^6\, 3s^2\, 3p^6\, 4s^2\, 3d^8$

7 There are two possible ways that an electron could move from the $4p^1$ sub-shell back to the $4s^1$ sub-shell: a) directly from $4p^1$ to $4s^1$ and b) from $4p^1$ to $3d^1$ then to $4s^1$. Each of these three transitions ($4p^1$ to $4s^1$; $4p^1$ to $3d^1$; $3d^1$ to $4s^1$) will emit a different photon.

8 The mass number of lithium-7 refers to the mass of the specific isotope, lithium-7. The relative atomic mass is the mean mass of all lithium atoms (taking into account isotopic abundance) relative to 1/12 of carbon-12.

9 a C_2H_4
  C: $2 \times 12 = 24$
  H: $4 \times 1 = 4$
  $M_r = 28$
  b NH_3
  N: $1 \times 14 = 14$
  H: $3 \times 1 = 3$
  $M_r = 17$
  c CCl_4
  C: $1 \times 12 = 12$
  Cl: $4 \times 35.5 = 142$
  $M_r = 154$

d C_3H_7OH
  C: $3 \times 12 = 36$
  H: $8 \times 1 = 8$
  O: $1 \times 16 = 16$
  $M_r = 60$

10 a Na_2O
  Na: $2 \times 23 = 46$
  O: $1 \times 16 = 16$
  RFM = 62
  b K_2SO_4
  K: $2 \times 39 = 78$
  S: $1 \times 32 = 32$
  O: $4 \times 16 = 64$
  RFM = 174
  c Al_2O_3
  Al: $2 \times 27 = 54$
  O: $3 \times 16 = 48$
  RFM = 102
  d $(NH_4)HCO_3$
  N: $1 \times 14 = 14$
  H: $1 \times 5 = 5$
  C: $1 \times 12 = 12$
  O: $3 \times 16 = 48$
  RFM = 79

11 Atomic mass involves neutrons as well as protons, and the number of neutrons in an atom can vary (as isotopes). The atomic (proton) number is unique to each element.

12

Group name	Group	Block	Period	Element
Alkali metals	1	s	2	Li
Alkaline earth metals	2	s	4	Ca
Transition metals	(8)	d	4	Fe
Halogens	7 (17)	p	3	Cl
Noble gases	0 (18)	p	4	Kr

13 a Be
  b Al
  c Sc
  d Kr

14 a Increasing the atomic number by 1, increases the Pauling electronegativity value by 0.5, starting from lithium (3,1.0) up to fluorine (9,4.0).
  b Neon is a noble gas and has full electron shells, so does not share electrons.

	Periodic Table group	Alkali metals	Alkaline earth metals	Halogens	Noble gases
How the property changes down the group	**Atomic radius**	**Increases**	**Increases**	Increases	**Increases**
	Ionisation energy	**Decreases**	Decreases	**Decreases**	**Decreases**
	Electronegativity	**Decreases**	**Decreases**	**Decreases**	Not applicable
	Reactivity	**Increases**	**Increases**	**Decreases**	Not applicable

16 a Formula of lithium = Li

Relative atomic mass of lithium, A_r = 6.9 g mol^{-1}

$$\text{moles} = \frac{\text{mass}}{A_r} = \frac{2.3\,\text{g}}{6.9\,\text{g mol}^{-1}} = 0.33\,\text{mol}\ (2\,\text{sf})$$

b Formula of salt = NaCl

Relative formula mass of sodium chloride, RFM = 23.0 + 35.5 = 58.5 g mol^{-1}

$$\text{moles} = \frac{\text{mass}}{\text{RFM}} = \frac{16\,\text{g}}{58.5\,\text{g mol}^{-1}} = 0.27\,\text{mol}\ (2\,\text{sf})$$

c Formula of calcium carbonate = $CaCO_3$

Relative formula mass of calcium carbonate, RFM = 40.1 + 12.0 + (16.0 × 3) = 100.1 g mol^{-1}

$$\text{moles} = \frac{\text{mass}}{\text{RFM}} = \frac{47\,\text{g}}{100.1\,\text{g mol}^{-1}} = 0.47\,\text{mol}\ (2\,\text{sf})$$

17 a Atomic mass, A_r, of sulfur = 32.1 g mol^{-1}

Mass of 2.5 moles = 32.1 g mol^{-1} × 2.5 mol = 80.25 g = 80 g (2 sf)

b Relative formula mass, RFM of magnesium oxide = 24.3 + 16.0 = 40.3 g mol^{-1}

Mass of 0.4 moles = 40.3 g mol^{-1} × 0.4 mol = 16.12 g = 16 g (2 sf)

c Relative formula mass, RFM, of sodium sulfate = (23.0 × 2) + 32.1 + (16.0 × 4) = 142.1 g mol^{-1}

Mass of 0.02 moles = 142.1 g mol^{-1} × 0.02 mol = 2.842 g = 2.8 g (2 sf)

Maths skills practice questions

1 $pV = nRT \Rightarrow V = \dfrac{nRT}{p}$

$$= \frac{2 \times 8.3145\,\text{J mol}^{-1}\,\text{K}^{-1} \times 273.15\,\text{K}}{1 \times 10^5\,\text{Pa}}$$

$$= 0.045\,\text{m}^3\ (2\,\text{sf})$$

2 $n = \dfrac{pV}{RT} = \dfrac{1 \times 10^5\,\text{Pa} \times 5.8\,\text{m}^3}{8.3145\,\text{J mol}^{-1}\,\text{K}^{-1} \times 298.15\,\text{K}}$

$= 233.97$ moles = 230 moles (2 sf)

Now test yourself

18 $50\,\text{cm}^3 = \dfrac{50\,\text{cm}^3}{1000} = 0.05\,\text{dm}^3$

Number of moles = volume (dm^3) × concentration (mol dm^{-3}) = 0.05 dm^3 × 0.12 mol dm^{-3} = 0.006 moles

19 $500\,\text{cm}^3 = \dfrac{500\,\text{cm}^3}{1000} = 0.5\,\text{dm}^3$

$$\text{concentration (mol dm}^{-3}) = \frac{\text{moles}}{\text{volume (dm}^3)} = \frac{1.8\,\text{moles}}{0.5\,\text{dm}^3}$$

$$= 3.6\,\text{mol dm}^{-3}$$

20 a $MgO(s) + 2HCl(aq) \rightarrow MgCl_2(aq) + H_2O(l)$

b $Na_2CO_3(s) + H_2SO_4(aq) \rightarrow Na_2SO_4(aq) + H_2O(l) + CO_2(g)$

c $C_3H_8(g) + 5O_2(g) \rightarrow 3CO_2(g) + 4H_2O(l)$

21

Element	nitrogen, N	oxygen, O
Mass used (g)	0.32 g	0.74
Atomic mass, A_r	14.0	16.0
Number of moles	$\dfrac{0.32\,\text{g}}{14.0\,\text{g mol}^{-1}} = 0.02286\,\text{mol}$	$\dfrac{0.74\,\text{g}}{16.0\,\text{g mol}^{-1}} = 0.04625\,\text{mol}$
Ratio of moles (÷0.02286)	1	2.023 ≈ 2
Empirical formula of magnesium oxide	NO_2	

22

Element	magnesium, Mg	chlorine, Cl
Mass used (g)	2.60	7.60
Atomic mass, A_r	24.3	35.5
Number of moles	$\dfrac{2.60\,\text{g}}{24.3\,\text{g mol}^{-1}} = 0.10700\,\text{mol}$	$\dfrac{7.60\,\text{g}}{35.5\,\text{g mol}^{-1}} = 0.21408\,\text{mol}$
Ratio of moles (÷0.10700)	1	2.0001 ≈ 2
Empirical formula of magnesium chloride	$MgCl_2$	

23 $Ca(s) + F_2(g) \rightarrow CaF_2(s)$

Atomic mass of calcium = $40.1\,g\,mol^{-1}$

Number of moles of calcium burning = $\dfrac{3.2\,g}{40.1\,g\,mol^{-1}}$
= 0.0798 moles = 0.080 moles (2 sf)

Number of moles of calcium fluoride formed = 0.080

Relative formula mass of CaF_2 = 40.1 + (2 × 19.0)
= $78.1\,g\,mol^{-1}$

Mass of CaF_2 formed = $78.1\,g\,mol^{-1}$ × 0.080 moles
= 6.248 g = 6.2 g (2 sf)

24 $ZnCO_3(s) \rightarrow ZnO(s) + CO_2(g)$

Relative formula mass of zinc carbonate
= 65.4 + 12.0 + (3 × 16.0) = $125.4\,g\,mol^{-1}$

Number of moles of zinc carbonate decomposing

= $\dfrac{15\,g}{125.4\,g\,mol^{-1}}$ = 0.12 moles (2 sf)

Number of moles of calcium fluoride formed
= 0.12 moles

Relative formula mass of CO_2 = 12.0 + (2 × 16.0)
= $44.0\,g\,mol^{-1}$

Mass of CO_2 formed = $44.0\,g\,mol^{-1}$ × 0.12 moles
= 5.28 g = 5.3 g (2 sf)

25 In a strong acid, most of the acid molecules dissociate, but as the acid is dilute, there are few acid molecules per dm^3. In a weak acid, few of the acid molecules dissociate, but as the acid is concentrated, there are many acid molecules per dm^3.

26 The equivalence point can be determined for a given volume of acid, by the volume of alkali required to change the colour of the indicator, OR, the volume of alkali added where the pH rises rapidly.

27 There is no sharp equivalence point, so an indicator will gradually change colour, making it difficult to determine the exact volume. It is better to use a pH probe to identify the equivalence point, then record the volume added to get a pH of 7.

28 Phenolphthalein

29 $HCl(aq) + LiOH(aq) \rightarrow LiCl(aq) + H_2O(l)$

Number of moles of HCl present

= $\dfrac{25.0\,cm^3}{1000}$ × $0.12\,mol\,dm^{-3}$ = 0.003 moles

At the equivalence point, the number of moles of LiOH used = 0.003 moles, so:
Concentration of LiOH =

$0.003\,moles \times \dfrac{1000}{28.9\,cm^3}$ = 0.1038
= $0.104\,mol\,dm^{-3}$ $\left(3\,sf\right)$

30 $H_2SO_4(aq) + 2KOH(aq) \rightarrow K_2SO_4(aq) + 2H_2O(l)$

The stoichiometry states that at neutralisation, 1 mole of H_2SO_4 has reacted with 2 moles of KOH. The acid reacts with the alkali in a ratio of 1:2.

Number of moles of H_2SO_4 present

= $\dfrac{25.0\,cm^3}{1000}$ × $3.5\,mol\,dm^{-3}$ = 0.0875 moles

At the equivalence point, the number of moles of NaOH used = 2 × 0.0875 = 0.175 moles, so:

Concentration of KOH =

$0.175\,moles \times \dfrac{1000}{15.6\,cm^3}$ = $11.2\,mol\,dm^{-3}$ $\left(3\,sf\right)$

31 a Li_2SO_4

b $Ca(NO_3)_2$

c $Fe_2(CO_3)_3$

d $Mg(OH)_2$

32 Magnesium chloride is an ionic compound. Electrical insulator when solid, conductor when molten or in aqueous solution. High melting and boiling points. Not volatile. Soluble in water, insoluble in non-polar solvents.

33 The bonds between the copper ions and the sulfate ions are too strong for water/non-polar solvents to break up the lattice.

34 Poor electrical conductor. Low melting and boiling points. Volatile. Insoluble in water and soluble in non-polar solvents.

35 The delocalised electrons between the layers are free to move between the layers (and so carry electrical charge between the layers) but do not move at right angles through the layers.

36 Graphene nanotubes are very strong, lightweight and flexible. They have high melting points and will conduct electricity down the nanotubes. The suits for racing drivers would be an ideal use for graphene nanotubes.

37 Aluminium is lightweight but strong. Electrical signals can travel through the frames and panels if required. It has a high melting and boiling point, so it will stay solid on the surface of Mars and during takeoff and landing.

38 In brass, zinc atoms take the place of some copper ions in the metallic structure. In steel, carbon atoms fill in some of the spaces between the iron atoms.

39 The enthalpy of reaction refers to the enthalpy changes for the formula:

$$2H_2(g) + O_2(g) \rightarrow 2H_2O(l)$$

The enthalpy of formation of water refers to the enthalpy changes for the reaction forming 1 mole of water, i.e.

$$H_2(g) + \frac{1}{2}O_2(g) \rightarrow H_2O(l)$$

The enthalpy of combustion of hydrogen refers to the combustion of 1 mole of hydrogen in an excess of oxygen – this will be the same as the enthalpy change for 1 mole of water.

40 Scale diagram, similar to Figure 1.33.

Activation energy (to scale) +2650 kJ mol⁻¹

Enthalpy of reaction (to scale) –890 kJ mol⁻¹

41

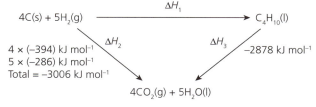

$$\Delta H_1 = \Delta_f H = \Delta H_2 - \Delta H_3$$
$$= -3006\,kJ\,mol^{-1} - (-2878)\,kJ\,mol^{-1}$$
$$= (-3006 + 2878)\,kJ\,mol^{-1}$$
$$= -128\,kJ\,mol^{-1}$$

42

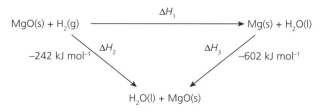

$$\Delta H_1 = \Delta_{rx} H = \Delta H_2 - \Delta H_3$$
$$= -242\,kJ\,mol^{-1} - (-602)\,kJ\,mol^{-1}$$
$$= (-242 + 602)\,kJ\,mol^{-1}$$
$$= +360\,kJ\,mol^{-1}$$

43

Bonds broken	Enthalpy /kJ mol⁻¹		Bonds made	Enthalpy /kJ mol⁻¹
1 × C=C	+619		4 × O–H	–(4 × 464) = –1856
4 × C–H	+(4 × 414) = +1656		4 × C=O	–(4 × 724) = –2896
3 × O=O	+(3 × 499) = +1497			
Total	+3772			–4752

$$\Delta_{rx} H = +3772\,kJ\,mol^{-1} - 4752\,kJ\,mol^{-1} = -980\,kJ\,mol^{-1}$$

44

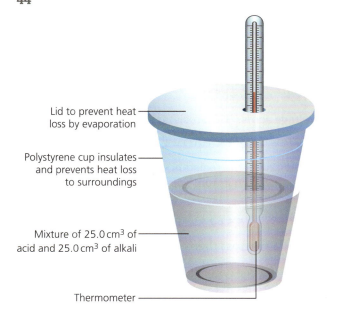

Lid to prevent heat loss by evaporation

Polystyrene cup insulates and prevents heat loss to surroundings

Mixture of 25.0 cm³ of acid and 25.0 cm³ of alkali

Thermometer

Physics

Maths skills practice questions

1 $2.6\,GW = 2.6 \times 10^9\,W$ – this is the *useful power output*

$32\,TJ$ per hour $= 32 \times 10^{12}\,J$ per hour

$$= \frac{32 \times 10^{12}\,J}{3600\,s} = 8.9 \times 10^9\,W$$ – this is the *total power input*

$$efficiency = \frac{useful\ power\ output}{total\ power\ input} \times 100\%$$

$$= \frac{2.6 \times 10^9\,W}{8.9 \times 10^9\,W} \times 100\% = 29\%$$

In this case, the data are given to 2 sf, so the answer must be given to 2 sf.

2 $$efficiency = \frac{useful\ energy\ output}{total\ energy\ input} \times 100\%$$

Rearranging the equation:

$$useful\ energy\ output = \frac{efficiency \times total\ energy\ input}{100\%}$$

$$= \frac{42\% \times 21.6\,J}{100\%} = 9.072\,J$$

$$= 9.1\,J\ (2\ sf)$$

3 The useful power output of the motor,

$$P = \frac{E}{t} = \frac{24\,J}{4\,s} = 6\,W$$

$$efficiency = \frac{useful\ power\ output}{total\ power\ input} \times 100\%$$

$$= \frac{6\,W}{15\,W} \times 100\% = 40\%$$

The least precise piece of data in the question is given to 1 sf, so the answer is to the correct sf.

Now test yourself

1 Because some energy will always be wasted ultimately as thermal energy by the device while it is working. Mechanical devices waste energy via friction and electrical devices waste energy due to resistive heating.

2 a 25% effective mobility

5% friction

30% coolant

40% exhaust gas

b 25% efficient

c Electric engines are much more efficient, so the effective mobility arrow will be much wider. Electric engines do not need a coolant and do not waste energy through exhaust gases, so these arrows will be absent.

d Surfboards are very efficient because they have a streamlined/hydrodynamic shape. Their bottom surface, which is in contact with the water, is also made from a very low-friction material.

3 Steel is very strong and it is a good conductor of heat, so it will easily transfer the thermal energy from the burning wood to the outside of the stove. The stove is painted black so that it easily radiates thermal energy into the surroundings.

4   Air is trapped between the layers of clothes. Air is a good insulator. This reduces heat loss through the clothes.

5   B

Maths skills practice questions

4   $U = \dfrac{Q}{At\Delta T}$, but $\dfrac{Q}{t} = P$, so:

$U = \dfrac{P}{A\Delta T} = \dfrac{250\,\text{W}}{6\,\text{m}^2 \times (20 - (-5))^\circ\text{C}}$

$= 1.666\,\text{W}\,\text{m}^{-2}\,^\circ\text{C}^{-1} = 1.7\,\text{W}\,\text{m}^{-2}\,^\circ\text{C}^{-1}\ (2\ \text{sf})$

5   $U = \dfrac{Q}{At\Delta T} \Rightarrow Q = UAt\Delta T$

$= 0.31\,\text{W}\,\text{m}^{-2}\,^\circ\text{C}^{-1} \times 8.4\,\text{m}^2 \times (6\,\text{h} \times 3600\,\text{s})$

$\times (19 - 3)^\circ\text{C}$

$= 899\,942.4\,\text{J} = 900\,000\,\text{J}$

6   Total U-value $= 5.7\,\text{W}\,\text{m}^{-2}\,^\circ\text{C}^{-1}$ and

$P = UA\Delta T = 5.7\,\text{W}\,\text{m}^{-2}\,^\circ\text{C}^{-1} \times (2\,\text{m} \times 3\,\text{m}) \times 18\,^\circ\text{C}$

$= 615.6\,\text{W} = 620\,\text{W}\ (2\ \text{sf})$

Now test yourself

6   B

7   Water is stored behind a dam, where it has an excess of gravitational potential energy.

   The water flows through pipes in the dam wall, gaining kinetic energy.

   The kinetic energy of the moving water turns a turbine, attached to a generator, which converts the kinetic energy into electricity, which is transferred to the National Grid.

8   HEP stations can only be built in places where there is a suitable river valley. This means most HEP stations are in highland areas, away from where the electricity is needed most. HEP stations need a large dam that destroys the habitat behind the dam.

9   a $I = \dfrac{V}{R} = \dfrac{6.0\,\text{V}}{150\,\Omega} = 0.040\,\text{A}$. Answer needs to be to 2 sf due to numbers in the data.

   b $I = \dfrac{Q}{t} \Rightarrow Q = It = 0.040\,\text{A} \times 3.0\,\text{min} \times 60\,\text{s}$

      $= 7.2\,\text{C}\ (2\ \text{sf})$

   c $P = VI = 6.0\,\text{V} \times 0.040\,\text{A} = 0.24\,\text{W}\ (2\ \text{sf})$

10 a   A series circuit consisting of a cell, a switch and a lamp.

   b $V = IR = 2.5 \times 10^{-3}\,\text{A} \times 1.8 \times 10^3\,\Omega = 4.5\,\text{V}\ (2\ \text{sf})$

   c $P = VI = 4.5\,\text{V} \times 2.5 \times 10^{-3}\,\text{A} = 0.01125\,\text{W}$

      $= 0.011\,\text{W}\ (2\ \text{sf})$

11 a $R = \dfrac{V}{I} = \dfrac{12\,\text{V}}{12.5\,\text{A}} = 0.96\,\Omega\ (2\ \text{sf})$

   b $P = VI = 12\,\text{V} \times 12.5\,\text{A} = 150\,\text{W}\ (2\ \text{sf})$

12 a $V = IR = 13\,\text{A} \times 8.9\,\Omega = 115.7\,\text{V} = 120\,\text{V}\ (2\ \text{sf})$

   b   Rate of heat loss

      $= I^2R = (13\,\text{A})^2 \times 8.9\,\Omega = 1504.1\,\text{W} = 1500\,\text{W}\ (2\ \text{sf})$

13 a   Rate of heat loss, $P = I^2R$

      $\Rightarrow I = \sqrt{\dfrac{P}{R}} = \sqrt{\dfrac{1.9 \times 10^3\,\text{W}}{150\,\Omega}} = 3.559\,\text{A} = 3.6\,\text{A}\ (2\ \text{sf})$

   b $V = IR = 3.6\,\text{A} \times 150\,\Omega = 540\,\text{V}\ (2\ \text{sf})$

14 a $R_T = R_1 + R_2 + R_3 = 75\,\Omega + 75\,\Omega + 75\,\Omega = 225\,\Omega$

   b $I_T = \dfrac{V}{R_T} = \dfrac{4.5\,\text{V}}{225\,\Omega} = 0.020\,\text{A}\ (2\ \text{sf})$

   c   Rate of heat loss

      $= I^2R = (0.020\,\text{A})^2 \times 75\,\Omega = 0.030\,\text{W}\ (2\ \text{sf})$

15 a   A parallel circuit containing a 12 V power supply, a lamp and two resistors.

   b $I_{12k\Omega} = \dfrac{V}{R} = \dfrac{12\,\text{V}}{12 \times 10^3\,\Omega} = 0.0010\,\text{A}$

      $I_{8k\Omega} = \dfrac{V}{R} = \dfrac{12\,\text{V}}{8.0 \times 10^3\,\Omega} = 0.0015\,\text{A}$

   c   Total current $= (0.0080 + 0.0010 + 0.0015)\,\text{A}$

      $= 0.0105\,\text{A} = 0.011\,\text{A}\ (2\ \text{sf})$

   d   Resistance of the lamp, $R_L = \dfrac{12\,V}{0.80 \times 10^{-3}\,A} = 15000\,\Omega$

      Total resistance of the circuit, R_T, is calculated by:

      $\dfrac{1}{R_T} = \dfrac{1}{15000\,\Omega} + \dfrac{1}{12000\,\Omega} + \dfrac{1}{8000\,\Omega} = 2.75 \times 10^{-4}\,\Omega^{-1}$

      $R_T = \dfrac{1}{2.75 \times 10^{-4}\,\Omega^{-1}} = 3636\,\Omega = 3600\ (2\,\text{sf})$

16 Maximum current will be drawn when the total resistance is lowest. 10.8 V battery voltage needs to be shared 6.1 V : 4.7 V, so the resistor chosen needs to be lower than 3.9 kΩ. The current being drawn, I, must be:

$I = \dfrac{V}{R} = \dfrac{6.1\,\text{V}}{3.9 \times 10^3\ \Omega} = 0.00156\,\text{A} = 0.0016\,\text{A}\ (2\ \text{sf})$

The current through both resistors will be the same as they are arranged in series, so:

$R = \dfrac{V}{I} = \dfrac{4.7\,\text{V}}{0.0016\,\text{A}} = 2937.5\,\Omega = 2900\,\Omega\ (2\ \text{sf})$

17 Minimum voltage occurs when the resistance is smallest (e.g. 75 Ω):

$I = \dfrac{V}{R_T} = \dfrac{18\,\text{V}}{(220 + 75)\ \Omega} = 0.061\,\text{A}\ (2\ \text{sf})$

Voltage across 75 Ω resistor is:

$V = IR = 0.061\,\text{A} \times 75\,\Omega = 4.575\,\text{V} = 4.6\,\text{V}\ (2\ \text{sf})$

Maximum voltage occurs when the resistance is biggest (e.g. 550 Ω):

$I = \dfrac{V}{R_T} = \dfrac{18\,\text{V}}{(220 + 550)\ \Omega} = 0.023\,\text{A}\ (2\ \text{sf})$

Voltage across 550 Ω resistor is:

$V = IR = 0.023\,\text{A} \times 550\,\Omega = 12.65\,\text{V} = 13\,\text{V}\ (2\ \text{sf})$

18 a   At 120 °C, the resistance of the thermistor is 2.2 kΩ from graph. Total current is therefore:

      $I = \dfrac{V}{R_T} = \dfrac{6.5\,\text{V}}{(5.5 + 2.2) \times 10^3\ \Omega} = 8.4 \times 10^{-4}\,\text{A}\ (2\ \text{sf})$

      Voltage across thermistor is:

      $V = IR = 8.4 \times 10^{-4}\,\text{A} \times 2200\,\Omega = 1.848\,\text{V}$

      $= 1.8\,\text{V}\ (2\ \text{sf})$

Check your understanding and progress at **www.hoddereducation.co.uk/myrevisionnotes**

b When voltage across thermistor = 3.2 V, the voltage across the 5.5 kΩ resistor = (6.5 – 3.2) V = 3.3 V

The current through both components is:

$$I = \frac{V}{R} = \frac{3.3\,\text{V}}{5.5 \times 10^3\,\Omega} = 6.0 \times 10^{-4}\,\text{A}\ (2\,\text{sf})$$

The resistance of the thermistor is given by:

$$R = \frac{V}{I} = \frac{3.2\,\text{V}}{6.0 \times 10^{-4}\,\text{A}} = 5333\,\Omega = 5.3\,\text{k}\Omega$$

From graph, temperature at 5.3 kΩ = 37 °C

19 Because it has a lot of free electrons that can easily flow through its structure.

20 Adding extra free electrons increases the conductivity and makes the semiconductor a better conductor of electricity.

21 The free electrons in a metal move constantly, in random directions. When the metal is heated, the mean speed of the free electrons increases. This means that they collide with themselves and the structure of the metal at a faster rate. The structure also vibrates at a higher rate and with a greater amplitude, also increasing the collision frequency. This increases the resistance of the metal.

22 a Resistance of R_1 is fixed, so taking a point, (0.02 A, 3.2 V):

$$R = \frac{V}{I} = \frac{3.2\,\text{V}}{0.02\,\text{A}} = 160\,\Omega\ (2\,\text{sf})$$

b Resistance of R_2 is fixed, so taking a point, (0.02 A, 6.4 V):

$$R = \frac{V}{I} = \frac{6.4\,\text{V}}{0.02\,\text{A}} = 320\,\Omega\ (2\,\text{sf})$$

c The resistance of the filament wire in the lamp increases with temperature, which increases with current.

d $R = \dfrac{V}{I} = \dfrac{8.0\,\text{V}}{0.02\,\text{A}} = 400\,\Omega$

23

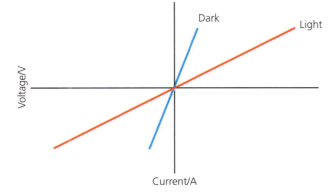

24 i Connect the thermistor in series with a (1 kΩ) standard resistor and a power supply.

ii Connect a voltmeter in parallel with the standard resistor.

iii Put the thermistor into a beaker of 80 °C hot water together with a thermometer.

iv Measure and record the temperature and the voltage.

v Add cold water to the beaker using a pipette until the temperature is 70 °C.

vi Measure and record the temperature and the voltage.

vii Repeat for temperatures of 60, 50, 40, 30, 20, 10 and 0 °C, using crushed ice to cool the water where required.

25 A large truck has more mass than a motorbike. Inertia depends on mass as well as velocity.

26 $F = ma = 54\,\text{kg} \times 3.2\,\text{m s}^{-2} = 172.8\,\text{N} = 170\,\text{N}\ (2\,\text{sf})$

27 The weight of the bungee jumper is equal and opposite to the reaction force/tension in the bungee cord.

28 $p = mv = 6.7 \times 10^{-3}\,\text{kg} \times 250\,\text{m s}^{-1}$

$$= 1.675\,\text{kg m s}^{-1}$$

$$= 1.7\,\text{kg m s}^{-1}\ (2\,\text{sf})$$

29 Total momentum before collision = total momentum after collision

$(0.15\,\text{kg} \times 0.75\,\text{m s}^{-1}) + 2.7 \times 10^{-3}\,\text{kg} \times 0\,\text{m s}^{-1} = (0.015\,\text{kg} \times 0\,\text{m s}^{-1}) + (2.7 \times 10^{-3}\,\text{kg} \times v_{\text{ball}})$

$$v_{\text{ball}} = \frac{(0.15\,\text{kg} \times 0.75\,\text{m s}^{-1})}{2.7 \times 10^{-3}\,\text{kg}}$$

$$= 41.66666\,\text{m s}^{-1} = 42\,\text{m s}^{-1}\ (2\,\text{sf})$$

30 Stretching the seatbelt material increases the time of the collision. This decreases the force on the driver/passengers.

31 $s = -$

$u = 9.5\,\text{m s}^{-1}$

$v = ?$

$a = 1.2\,\text{m s}^{-2}$

$t = 2.5\,\text{s}$

Equation needed:

$$v = u + at = 9.5\,\text{m s}^{-1} + \left(1.2\,\text{m s}^{-2} \times 2.5\,\text{s}\right) = 12.5\,\text{m s}^{-1}$$

32 $s = 50\,\text{m}$

$u = 12\,\text{m s}^{-1}$

$v = ?$

$a = 0.85\,\text{m s}^{-2}$

$t = -$

Equation needed:

$$v^2 = u^2 + 2as \Rightarrow v = \sqrt{u^2 + 2as}$$

$$= \sqrt{(12\,\text{m s}^{-1})^2 + 2 \times 0.85\,\text{m s}^{-2} \times 50\,\text{m}} = 15.1327\,\text{m s}^{-1}$$

$$= 15\,\text{m s}^{-1}\ (2\,\text{sf})$$

33 $s = ?$

$u = 0\,\text{m s}^{-1}$

$v = -$

$a = 9.8\,\text{m s}^{-2}$

$t = 2.7\,\text{s}$

Equation needed: $s = ut + \dfrac{1}{2}at^2$

The bungee jumper is initially at rest so $u = 0\,\text{m s}^{-1}$ and the equation becomes:

$$s = \frac{1}{2}at^2 = \frac{1}{2} \times 9.8\,\text{m s}^{-2} \times (2.7\,\text{s})^2 = 35.721\,\text{m}$$

$$= 36\,\text{m}\ (2\,\text{sf})$$

34 a Stationary, 1 m away from the origin, for 1 second.

b Travelling at a constant velocity of 3.0 m s⁻¹ for 2 seconds, away from the origin.

c Stationary, 7 m away from the origin, for 2 seconds.

d Travelling at a constant velocity of 2.0 m s⁻¹ for 2 seconds, towards the origin.

e Stationary, 3 m away from the origin, for 1 second.

f Travelling at a constant velocity of 1.5 m s⁻¹ for 2 seconds, towards the origin.

35 a $\text{acceleration} = \dfrac{\text{change in velocity}}{\text{time taken}} = \dfrac{9\ \text{m s}^{-1}}{6\ \text{s}}$

$= 1.5\ \text{m s}^{-2}$

b Distance travelled = area under graph

0–6 s, distance travelled $= \dfrac{1}{2} \times 6\ \text{s} \times 9\ \text{m s}^{-1} = 27\ \text{m}$

6–10 s, distance travelled $= 4\ \text{s} \times 9\ \text{m s}^{-1} = 36\ \text{m}$

Total distance travelled $= 27\ \text{m} + 36\ \text{m} = 63\ \text{m}$

Maths skills practice questions

7 $KE = \dfrac{1}{2}mv^2 = \dfrac{1}{2} \times 1.8 \times 10^3\ \text{kg} \times \left(26\ \text{m s}^{-1}\right)^2$

$= 608\,400\ \text{J}$

$= 610\,000\ \text{J}\ \left(2\ \text{sf}\right)$

8 $GPE = mgh = 430 \times 10^{-3}\ \text{kg} \times 9.8\ \text{N kg}^{-1} \times 14\ \text{m}$

$= 58.996\ \text{J} = 59\ \text{J}\ \left(2\ \text{sf}\right)$

9 $P = \dfrac{E}{t} \Rightarrow E = Pt = 1500\ \text{W} \times 5\ \text{minutes} \times 60\ \text{s}$

$= 450\,000\ \text{J}$

Unit 3   REVISED ●

Now test yourself

1 A source is a report/article about a subject.

2 Specialist scientific publishers / scientific blogs / broadsheet newspapers / tabloid newspapers / popular blogs [any 4]

3 An article in a scientific journal will have been peer-reviewed, whereas an article in a tabloid newspaper will not.

4 The RMS is likely to present views derived from climate change facts based on data; this information will be peer-reviewed. FotE is likely to present a biased view and may not include information that is counter to their views; this information may or may not be peer-reviewed.

5 a Increase in ice sheet shrinkage = (532 – 255) billion tonnes = 277 billion tonnes

b % increase $= \left(\dfrac{277}{255}\right) \times 100\% = 109\%$

6 Scientific method consists of the following steps: observation (of a phenomenon); question (identify the problem); research (search for existing solutions); hypothesise (come up with a hypothesis); experiment (design and carry out an experiment to collect data); test (analyse the data to accept or reject the hypothesis); conclusion (make a conclusion related to the hypothesis); report (report results).

7 Government (through research councils in the UK) / charities / commercial companies

8 The peer-review process consists of: a scientist or researcher submits an article/paper to a scientific journal; the journal sends the article/paper to an anonymous reviewer who is qualified in the same scientific field; the reviewer comments on or checks the article/paper; the commented/corrected article/paper is amended by the scientist/researcher or the journal approves the article/paper without changes; the whole cycle is repeated as necessary.

9 Other scientists (working in the same field) had not corroborated the Russian team's data or checked its accuracy. There may be side effects that are potentially harmful, or the benefits/efficacy of the vaccine may not be as stated.

10 Any two from:

from experiments set up in the field and monitored *in situ* or remotely

by direct observation of locust swarms in the field via remote sensing from altitude or by satellites remotely by population tagging of the locusts

11 Evidence-based argumentation is the process that scientists go through when they disagree about scientific explanations (or claims). They use empirical data (or evidence) to justify their side of the argument (or rationale). This process can guide the work of scientists who identify weaknesses and limitations in other scientists' arguments, with a view to refining and improving scientific explanations and the way that experiments are designed.

12 The trial was too small and could not have been a 'double-blind' trial (as there were too few subjects), so the conclusions drawn were unreliable and misleading. The claims were picked up by the popular mainstream media and presented as fact. This led to a reduction in the number of people being vaccinated and a subsequent increase in cases of measles, mumps and rubella.

13 Benefits (any 2): reliable / uninterrupted source of electricity; large quantities of electricity can be produced; no CO_2 emissions during generation of electricity; jobs.

Drawbacks (any 2): cost of build; cost of electricity; potential for contamination / nuclear accident; CO_2 emissions during construction; decommissioning costs; decommissioning contamination.

14 Any one of:

loss of electricity supply causing problems for domestic dwellings / services / transport / business / education / care / medical systems etc.

potential for public disorder

potential for crime

loss of communications systems.

15 Any two of:

promoting hydroxychloroquine as a 'miracle' treatment

widespread scepticism following the disproval of the study

Check your understanding and progress at **www.hoddereducation.co.uk/myrevisionnotes**

ridicule of the scientists involved in the original study

ridicule of the 'personalities' promoting hydroxychloroquine as a miracle treatment.

16 Any example of Greenpeace's direct action such as: disruption of whaling fleets; occupation of North Sea oil rigs; gate-crashing high profile meetings; placing large boulders on the seafloor to deter large-scale fishing.

17 a Materials scientist – perform experiments to determine the properties of graphene and the combination of graphene with other materials involved with the construction of the helmets.

   b Product/process developer – produce the design for the helmet, incorporating the graphene elements.

   c Sport and exercise scientist – perform experiments to determine the properties needed by the helmet, i.e. strength; flexibility; comfort; ability to play in the helmet etc.

18 a Biologist – perform experiments to determine the effects of pollutants on freshwater fauna.

   b Ecologist – perform fieldwork studies of the river; collect and record water samples; collect and record numbers of different fauna/flora.

   c Toxicologist – perform experiments to determine the types of pollutants and the concentration of pollutants.

Unit 4 REVISED

Now test yourself

1 By contraction of (circular) muscles behind the food / by peristalsis

2 Mechanical digestion breaks up the food increasing the surface area for enzymes to work.

3 Water and proteins / peptides

4 The small intestine

5 Glycerol and fatty acids

6 The breakdown of large droplets into smaller ones

7 To absorb fatty acids and glycerol

8 Sodium is needed for the absorption of glucose, and glucose is needed for energy (by respiration).

9 Gastrin travels in the bloodstream, not through the gut.

10 Iron is needed to make red blood cells. A pregnant woman needs enough iron to make new blood cells in both herself and her baby.

11 Rickets is a disease where the bones are soft and bend. This is because lack of vitamin D means that the cartilage cannot absorb calcium to become rigid bone. In adults, bone has already formed.

12 Ribs / pelvis / skull / backbone

13 Osteoblasts

14 The skeleton will weaken / soften because calcium is needed to harden bone.

15 In old age, people tend to do less exercise and this leads to the removal of bone tissue / resorption.

16 Actin and myosin

17 It detached the myosin heads from actin.

18 It changes shape and removes the tropomyosin blocking the attachment sites.

19 Slow-twitch. Slow-twitch muscles are used for long-term contraction which is needed to maintain posture.

20 Myoglobin is an oxygen store used when there is not enough oxygen in the cell during intense activity. Slow-twitch muscle fibres are not involved in intense activity.

21 Lifting heavy weights over short periods of time promotes the development of fast-twitch muscles, needed by sprinters. Lifting lighter weights over a longer period promotes the development of slow-twitch fibres, needed for endurance in long-distance runners.

22 Plasma is mostly water and oxygen does not dissolve well in water.

23 Active tissues will be using oxygen for respiration, so the partial pressure of oxygen will be low. At low partial pressures, haemoglobin can only have a low % saturation, so to reach that it must release oxygen.

24 The Bohr effect is the shifting of the oxygen dissociation curve to the right at increased CO_2 concentrations, so reducing the affinity of haemoglobin for oxygen.

25 The extra red blood cells / haemoglobin made during high-altitude training persist for 10–14 days. The athletes' blood can therefore carry more oxygen for aerobic respiration.

26 It emits two beams of light of different wavelengths. Oxygen in the blood absorbs these two wavelengths to different extents. The pulse oximeter detects the difference in the absorption and converts this into a read-out of oxygen saturation.

27 Low blood pressure will decrease the rate of blood flow (and therefore the supply of oxygen) to muscles and other tissues. Those tissues get their energy from aerobic respiration which requires oxygen, so with less oxygen they will have less energy.

28 Parietal, occipital, temporal, frontal

29 a Peripheral

   b It will decrease the heart rate.

30 The frontal lobe of the cerebral cortex

31 The parietal lobe of the cerebral cortex because it controls orientation and recognition, or (possibly) the occipital lobe of the frontal cortex because it controls visual processing.

32 Away from the central nervous system

33 They move down a concentration gradient

34 Potassium / K^+

35 It causes sodium channels to open

36 Calcium ions are required at the pre-synaptic knob to cause the release of neurotransmitter.

37 Alzheimer's patients have low levels of acetylcholine. Acetylcholinesterase catalyses the breakdown of acetylcholine, so if it is inhibited acetylcholine levels will rise.

Index

Check your understanding and progress at **www.hoddereducation.co.uk/myrevisionnotes**

Index

201

My Revision Notes: AQA Applied Science Suitable for Level 3 and Level 3 Extended Certificates

Check your understanding and progress at **www.hoddereducation.co.uk/myrevisionnotes**

My Revision Notes: AQA Applied Science Suitable for Level 3 and Level 3 Extended Certificates

Check your understanding and progress at **www.hoddereducation.co.uk/myrevisionnotes**

My Revision Notes: AQA Applied Science Suitable for Level 3 and Level 3 Extended Certificates